Additive Manufacturing for Environmental Sustainability

The text begins by discussing the design, development, and advancements of additive manufacturing techniques. It then covers additive manufacturing for micro and nanomanufacturing, materials efficiency, and waste minimization. This book illustrates the processing of biomaterials, smart materials, and functionally graded materials for additive manufacturing.

Key Features:

- Explores the applications of additive manufacturing in promoting environmental sustainability, focusing on its role in waste reduction, resource conservation, energy efficiency, and emissions reduction.
- Illustrates the processing of biomaterials, smart materials, and functionally graded materials for additive manufacturing.
- Explains surface engineering and wear modeling of tribological properties for additive manufacturing.
- Discusses the challenges and opportunities associated with integrating additive manufacturing into sustainable practices and future directions for research and development in this burgeoning field.
- Showcases applications of additive manufacturing in the aerospace industry, automotive industry, and healthcare sector.

It is primarily written for senior undergraduates, graduate students, and academic researchers in manufacturing engineering, mechanical engineering, industrial engineering, aerospace engineering, production engineering, and automotive engineering.

Additive Manufacturing for Environmental Sustainability

Technology and Applications

Edited by

Ankit Tyagi,
Rekha Nair and
B. P. Naveen

CRC Press
Taylor & Francis Group
Boca Raton London New York

CRC Press is an imprint of the
Taylor & Francis Group, an **informa** business

First edition published 2027
by CRC Press
2385 NW Executive Center Drive, Suite 320, Boca Raton FL 33431

and by CRC Press
4 Park Square, Milton Park, Abingdon, Oxon, OX14 4RN

CRC Press is an imprint of Taylor & Francis Group, LLC

ISBN: 9781032999883 (hbk)
ISBN: 9781032999890 (pbk)
ISBN: 9781003607052 (ebk)

DOI: 10.1201/9781003607052

Typeset in Sabon
by codeMantra

Contents

Preface

Additive manufacturing (AM), commonly known as 3D printing, has emerged as a disruptive technology with vast potential to revolutionize traditional manufacturing processes. Beyond its efficiency and versatility in prototyping and production, AM holds promise for advancing environmental sustainability across various industries. This book aims to provide a comprehensive overview of how AM can contribute to environmental sustainability efforts, offering insights for researchers, practitioners, and policymakers interested in leveraging AM technologies for a greener and more sustainable future. This book explores the applications of AM in promoting environmental sustainability, focusing on its role in waste reduction, resource conservation, energy efficiency, and emissions reduction. Additionally, it discusses the challenges and opportunities associated with integrating AM into sustainable practices, as well as future directions for research and development in this burgeoning field. Conventional manufacturing relies on subtractive methods, whereas nature achieves desired shapes through additive processes. AM is revolutionizing design approaches and production processes, offering solutions for diverse applications, including biomedical objects like human tissue. Its disruptive impact extends to leveraging AI and Industry 4.0 advancements, transforming manufacturing capabilities and global markets across industries such as aerospace, automotive, life sciences, and defense.

This novel production method enables cost-effective realization of superior designs, enhancing productivity and sustainability. By swiftly converting 3D CAD data into physical prototypes, AM significantly reduces product development cycles. The rapid evolution of AM is the focus of this book, comprising scholarly contributions from researchers, academics, and industry professionals. It covers a spectrum of topics, including materials advancements, bio-fabrication, AI and Industry 4.0 integration, and case studies. Detailed discussions on process optimization and control challenges offer insights into AM's potential. Emerging developments like hybrid AM techniques for challenging materials, nature-inspired AM applications, and optimization through bio-inspired algorithms are explored. This book also delves into AI integration throughout the manufacturing process, including real-time defect detection and process efficiency enhancement.

Acknowledgments

The successful completion of this book, *Additive Manufacturing for Environmental Sustainability*, would not have been possible without the support, encouragement, and contributions of many individuals and institutions.

First and foremost, we express our sincere gratitude to all the researchers, academicians, and industry professionals whose pioneering work in additive manufacturing and sustainability has inspired the foundation of this book. Their invaluable contributions to knowledge and innovation have significantly shaped the content presented here.

We are deeply thankful to our colleagues and peers for their continuous support, insightful discussions, and constructive feedback throughout the development of this manuscript. Their perspectives have greatly enhanced the quality and depth of this work.

We extend our heartfelt appreciation to the institutions and organizations that provided the necessary resources, research environment, and motivation to pursue this endeavor. Their commitment to advancing sustainable technologies has been a constant source of inspiration.

Special thanks are due to the editorial and publishing team for their professionalism, patience, and guidance during the entire publication process. Their efforts have been instrumental in bringing this book to fruition.

We are profoundly grateful to our family for their unwavering support, understanding, and encouragement. Their belief in our work has been a driving force behind the completion of this book.

Finally, we acknowledge all the readers, students, and practitioners who are committed to advancing sustainable manufacturing practices. It is our hope that this book contributes meaningfully to the growing body of knowledge in additive manufacturing and supports the global pursuit of environmental sustainability.

About the editors

Ankit Tyagi is an associate professor in the Department of Mechanical Engineering, Poornima College of Engineering, Jaipur, India. He earned his Ph.D. from Delhi Technological University, India, in Manufacturing Processes in Advanced Coating. He has visited various academic institutions and industries in several countries for academic pursuits.

His area of research includes surface modification, coating, tribology, material processing, composite materials, metal matrix nanocomposites, nano-coatings, wear, deformation, and corrosion. He also has a good command of teaching subjects such as welding technology, casting, wear mechanics, materials science, non-destructive testing, metrology, and composite materials. He has published over 35 research papers in reputed journals and conference proceedings.

Rekha Nair is a chemistry professor and Dean of Academics at Poornima College of Engineering, Jaipur, Rajasthan, India. She earned her Ph.D. from the University of Rajasthan, Jaipur, in Corrosion Inhibition for Aluminium and Its Alloys. She is the chairperson of the Internal Complaint Committee (ICC) and Women in Science and Engineering (WISE) at Poornima College of Engineering. She has organized a number of workshops, conferences, and training at national and international levels. She has acted as the convenor for a number of conferences, including the International Conference on Sustainable Energy, Environment and Green Technologies, 2021, with publication support from Scopus-indexed *IOP Conference Series: Earth and Environmental Science (795)*. She has also acted as the convenor for the *Conference Series: International Conference on Recent Advances in Engineering (ICRAE)* since 2022. Her research areas are corrosion inhibition for metals, surface coatings, electrochemistry, and environmental pollution. She has over two decades of experience in teaching and administration with more than 50 publications in national/international journals and conferences. Six books authored by her are referred by several universities as textbooks. Her major works published are in Engineering Chemistry, Environmental Engineering and Disaster Management, and Chemistry and Environmental Engineering. She has authored many books of international repute. Her YouTube Channel on Engineering Chemistry, which has more than 50 lectures, helped students during COVID and has become popular among students across the world. She has received the "Excellence in Academic Leadership Award" recently from the "International EARG Awards 2024" for her contributions. She has contributed as a guest of honour for the "Three days residential awareness workshop on plastic waste management" at Central Institute of Plastics Engineering and Technology (CIPET), March 24–26, 2022. She has been a member of the Selection Committee for the reputed KVPY scholarship exam (Kishore Vaigyanik Protsahan Yojana exam) organized at the national level by IISc Bengaluru. She is a member of the Board of Studies at the Department of Chemistry, Parishkar College of Global Excellence (Autonomous) (PCGEA), Jaipur. She has acted as a member of the Planning and Monitoring Board, Poornima University, Jaipur; a member of the Advisory Board at Poornima College of Engineering; and a member of the Governing Council at Institutions under Poornima Group. She is a lifetime member of the Indian Society for Technical Education (ISTE). As an Institutional Coordinator for ISTE, she organizes many technical papers and model contests for students at the national level. She has been an invited speaker and has chaired the sessions at national and international conferences.

Her recent contributions include a book titled *Recent Advances in Green Technologies and Sustainable Development*, Taylor & Francis Group, 2023, and a copyright on "A Demand Aggregation Approach for Inventory Control in Two Echelon Supply Chain for Multi-Products", L-124614/2023, India, 2023. She has been the convener for Amrit Kaal Vimarsh Vikasit Bharat @2047 on the topic titled "Science, Technology and Innovation in the New Millennium: Brave New World", December 2023.

B. P. Naveen is a professor in the Department of Civil Engineering, National Institute of Technical Teachers' Training and Research (NITTTR), Ministry of Education, Government of India. He earned his B.E. from the University Visvesvaraya College of Engineering (UVCE), Bangalore; M.Tech (Geotechnical Engineering) from the National Institute of Technology Karnataka, Surathkal; and M.S. (Geotechnical Engineering), Ph.D., and Postdoctoral (Geotechnical Engineering) degrees from IISc, Bangalore. He was also involved in teaching, research, and administration. He has seven years of post-Ph.D. teaching experience, five years of industry experience after graduation and post-graduation, and six years of research experience during his Ph.D., amounting to a total experience of 18 years. His area of research is Geotechnical and Geoenvironmental Engineering. His Ph.D. work was in Assessment of Leachate Characteristics & Geotechnical Properties of Municipal Solid Waste Landfill.

His Ph.D. thesis is referred to by the Bruhat Bengaluru Mahanagara Palike (BBMP) and the Karnataka State Pollution Control Board for reopening the Mavallipura landfill. He served as a resource person and taught regular Ph.D. students on the topic entitled "Solid Waste Management" at the Centre for Nano and Soft Material Science (CeNS) research institute under the Department of Science and Technology (DST), Government of India, and NLD Training Institute, Bhutan.

He has 1,000 Google Scholar-cited international/national journal publications in various refereed journals like *ASTM Geotechnical Testing*, *Environmental Pollution*, and *Measurement*, including peer-reviewed conference proceeding papers. Out of these publications, 30 documents are SCI/Scopus indexed. He is the chief editor of an international journal in his area of research. During his academic career, he also collaborated on research with different groups and research centers. He has bagged six awards (SERB TARE, IEI Young Engineers Award, Professional Excellence Award, ISSMGE Foundation Award, Centre for International Co-operation in Science (CICS), and Sir M. Visvesvaraya Award) in the year 2019, and he was nominated and selected by the Indian Geotechnical Society (IGS) as one of the two Young Geotechnical Scientists (aged below 35 years) to represent India and present a research paper at the 9th Young Asian Geotechnical Engineers Conference (9AYGEC) and 15th International Conference on Geotechnical Engineering (15ICGE). He is a member of the Institute of Electrical and Electronics Engineers (IEEE), American Society of Civil Engineering (ASCE), American Society for Testing and Materials (ASTM), International Society of Soil Mechanics and Geotechnical Engineering (ISSMGE), Indian Geotechnical Society (IGS), Karnataka Environmental Research Foundation (KERF), National Foundation for Entrepreneurship Development (NFED), Consulting Engineers Association of India (CEAI), Institution of Engineering and Technology (IET), and Institution of Engineers India (IEI).

He has developed a Pile Modelling Design Kit using PLAXIS 2D and 3D, with detailed cases from the industry for fresh/intermediate Geotechnical Design Engineers. He is well versed in numerical packages (applicable in the Geotechnical and Geoenvironmental profession) like PLAXIS 2D and 3D, Pollute, Deep soil, Shake 2000, Geo 5, Alp Oasis, and AutoCAD. In addition, he handled the Soil Mechanics and Foundation Engineering course at the International Geotechnical Centre CGI, Lima, Peru.

He has guided several Ph.D., M.S., M.Tech, and B.Tech students. He has delivered over 65 invited lectures in various engineering, research institutions, and technical associations. He has handled the National Assessment and Accreditation Council (NAAC) and Sustainable Development Goals (SDG) accreditation process at Amity University.

He has also contributed to various consultancy projects during his stay in Amity University Haryana, such as the Surat Metro Project, the

Conversion of Meter Gauge Line to Dual Gauge from Parbatipur to Kaunia Section of Bangladesh Railway Project of Bangladesh Railway, and Installation and Commissioning of Handpumps in Barmer and Jaisalmer Districts. He has overseas work experience, having worked in Bangladesh, Kenya, Japan, and the USA.

Contributors

Akash Ahalwat
Department of Mechanical Engineering
UIET, Maharshi Dayanand University
Rohtak, Haryana, India

Deepanshu Arora
Department of Commerce and Management Faculty of Management Studies
WISDOM Banasthali Vidyapith
Niwai, Tonk, Rajasthan, India

Anant Bhardwaj
Department of Mechanical Engineering
Poornima College of Engineering
Jaipur, Rajasthan, India

Kandpal Bhaskar
Department of Mechanical Engineering
Indrapresth Engineirng College
Ghaziabad, India

Neeraj Bisht
Department of Mechanical Engineering
College of Agriculture
G.B. Pant University of Agriculture and Technology
Pantnagar, Uttarakhand, India

Manjeet Chahar
Department of Mechanical Engineering
Deenbandhu Chhotu Ram University of Science and Technology
Sonipat, Haryana, India

Arun Kumar Chaudhary
Industrial and Production Engineering Department
G.B. Pant University of Agriculture and Technology
Pantnagar, Uttarakhand, India

Swasti Chaudhary
Department of Mechanical & Automation Engineering
School of Automation
Banasthali Vidyapith
Rajasthan, India

Abhishek Dadhich
Department of Computer Science Engineering
Poornima Institute of Engineering and Technology
Jaipur, Rajasthan, India

Ankita Dadhich
Department of Computer Science & Engineering
Global Institute of Technology
Jaipur, Rajasthan, India

Deepak Dahiya
Centre of Excellence for Energy and Environmental Studies
Deenbandhu Chhotu Ram University of Science and Technology
Sonipat, Haryana, India

Ramesh Kumar Garg
Department of Mechanical Engineering
Deenbandhu Chhotu Ram University of Science and Technology
Sonipat, Haryana, India

Deepak Gupta
Department of Mechanical Engineering
Graphic Era Hill University
Dehradun, Uttarakhand, India

Kiran Kaushik
Centre of Excellence for Energy and Environmental Studies
Deenbandhu Chhotu Ram University of Science and Technology
Sonipat, Haryana, India

Shahroz Akhtar Khan
Department of Computer Science Engineering
Galgotias College of Engineering & Technology
Greater Noida, Uttar Pradesh, India

Amit Mandal
Department of Mechanical Engineering
Poornima College of Engineering
Jaipur, Rajasthan, India

Dipesh Kumar Mishra
Department of Mechanical Engineering
Graphic Era Deemed to be University
Dehradun, Uttarakhand, India

Vikas Misra
Mechanical Engineering Department
Shree Ramswaroop Memorial University
Barabanki, Uttar Pradesh, India

Rekha Nair
Department of Chemistry
Poornima College of Engineering
Jaipur, Rajasthan, India

Devesh Pant
Industrial and Production Engineering Department
G.B. Pant University of Agriculture and Technology
Pantnagar, Uttarakhand, India

Upender Punia
Department of Mechanical Engineering
Deenbandhu Chhotu Ram University of Science and Technology
Sonipat, Haryana, India

Nisha Rani
Department of Mechanical Engineering
G B Pant University of Agriculture and Technology
Pantnagar, Uttarakhand, India

Rohit Sahu
Department of Mechanical Engineering
G.L Bajaj Institute of Technology and Management
Greater Noida, Uttar Pradesh, India

Saurabh Sharma
Forace Polymer Pvt. Ltd.
Haridwar, Uttarakhand, India

Siddharth
Department of Civil Engineering
Poornima College of Engineering
Jaipur, Rajasthan, India

Parveen Sihmar
Department of Mechanical Engineering
Deenbandhu Chhotu Ram University of Science and Technology
Sonipat, Haryana, India

Dalvir Singh
Department of Mechanical Engineering
Galgotias College of Engineering & Technology
Greater Noida, Uttar Pradesh, India

Sachin Sirohi
Department of Mechanical Engineering, Faculty of Engineering and Technology
SRM Institute of Science and Technology
Delhi NCR Campus, Modinagar, India

Shreejaya Sivadas
Department of Post-Harvest Process and Food Engineering
G.B. Pant University of Agriculture and Technology
Pantnagar, Uttarakhand, India

Ankit Tyagi
Department of Mechanical Engineering
Poornima College of Engineering
Jaipur, Rajasthan, India

Alka Verma
Department of Mechanical Engineering
College of Agriculture
G. B. Pant University of Agriculture and Technology
Pantnagar, Uttarakhand, India

Brijesh Verma
Department of Mechanical Engineering
Galgotias College of Engineering & Technology
Greater Noida, Uttar Pradesh, India

Hemant Yadav
Department of Mechanical Engineering
Roorkee Institute of Technology
Roorkee, Uttarakhand, India

Chapter 1

Additive manufacturing in the field of multi-material and multi-functional printing

Brijesh Verma, Dalvir Singh, and Kandpal Bhaskar

1.1 INTRODUCTION

From a prototyping technology to a real manufacturing process capable of generating end-use components, additive manufacturing (AM) has evolved. Usually able only for single-material printing, conventional AM techniques limit the part functionality and complexity. But the development of multi-material and multi-functional printing systems has brought fresh possibilities by enabling the combination of several materials with different qualities and purposes into one construction. In multi-material printing, two or more materials are either concurrently or sequentially deposited during the printing process. This function helps parts with graded properties which is, varying stiffness, conductivity, or thermal resistance, to develop. This technology can perform multi-functional printing, which allows the printing of sensors, actuators, or even electronic circuits alongside components [1,2].

1.2 TECHNOLOGICAL ADVANCEMENTS IN MULTI-MATERIAL AND MULTI-FUNCTIONAL PRINTING

Advances in technology for printing with multiple materials and functions are changing how we make things. They offer unmatched accuracy and flexibility. These breakthroughs allow us to combine different materials and purposes in one print job, opening up new possibilities in design and use [3].

1.2.1 Multi-material printing technologies

Another aspect of AM is multi-material printing, which allows two or more materials to combine in a single build process. This functionality enables the design of components that may combine mechanical, thermal, electrical, or cosmetic performance factors combined into a single component. A number

DOI: 10.1201/9781003607052-1

of AM processes have been adapted or designed for multi-material printing. These include simple, accurate, and material-tolerant, and universal technologies such as:

1.2.1.1 Fused deposition modeling (FDM)

Fused deposition modeling (FDM), or fused filament fabrication (FFF), represents one of the most common techniques used in AM. FDM is known to be versatile and cost- efficient, while multi-material additive manufacturing enables the use of an ever-growing range of materials. For these reasons, it is capable of being utilized in the production of complex multi-functional parts which have distinct physical, mechanical, or even aesthetic requirements in a single print.

1.2.1.1.1 Working principle of FDM

In conventional FDM, the thermoplastic filament is put through a feeding system to a nozzle where it is melted and then extruded layer- by- layer to create a three-dimensional (3D) object. For multi-material FDM, this process can be expanded with the addition of multiple nozzles or a single mixing nozzle with many inputs. These configurations allow for the sequential or simultaneous printing of two or more materials.

1.2.1.1.2 Material combinations in FDM

FDM systems can work with a wide range of materials, which allows for numerous combinations in a multi-material configuration. Some of these materials include the following:

- **Rigid + Flexible Materials:** Such as PLA in combination with TPU for structural and flexible parts.
- **Structural + Support Materials:** Like ABS mixed with PVA (water-soluble) for removing support structures after printing complex geometries.
- **Conductive + Insulating Materials:** Used for the integration of basic circuits, antennas, or capacitive sensors into the 3D printed models.
- **Aesthetic Combinations:** Multiple colors or textures in a single print for artistic or branding purposes.

1.2.1.1.3 Advantages of multi-material FDM

- FDM enables the creation of hybrid parts that have different functionality areas (e.g., soft-touch areas, embedded joints, or variable color regions).
- It eliminates requiring refinement to an assembly by placing components like hinges and seals within a singular print.

- It makes the functioning of complex products possible in the real world to be rapidly prototyped.
- Generally, more affordable and accessible than other multi-material methods, FDM systems have better availability.
- It simplifies the cleanup process for internal cavities and overhangs, particularly with dissolvable supports.

1.2.1.2 Material jetting (MJ)

Material jetting (MJ) is a sophisticated form of additive manufacturing (AM) that applies inkjet printing methods to build 3D parts by precisely applying material in the form of micro droplets. Unlike other AM processes that usually construct objects from a single material, MJ has the ability to jet several materials at once, or in a specific order. This capability enables the fabrication of multi-material constituents with heterogenous mechanical, optical, and functional attributes. It is predominantly used in sectors where realism, precision, and functional testing are required, such as in automotive, electronics, consumer goods, and even healthcare. Its ability to print full-color models, transparent features, and parts with soft and rigid regions makes it one of the most versatile and precise AM technologies available.

1.2.1.2.1 Working principle of MJ

Material jetting operates like traditional 2D inkjet printing, but in three dimensions. In this process, multiple droplets of photopolymer resin are jetted from the nozzles of printheads against a build platform. The droplets are immediately cured using ultraviolet (UV) light to harden each layer before the next layer is jetted. In multiple material systems, a number of printheads, or print nozzles, jet simultaneously while dispensing a different material from each head or nozzle. The printer's software, which is the user's toolbox, manages the placement of the droplets, the size of the droplets, the sequential layering, and manages different materials into different patterns across specific regions. A support material is also printed in this process to support any overhangs, or complex geometric structures, and is later removed, generally through washing or peeling. The deposition continues layer- by- layer until a full 3D object is built.

1.2.1.2.2 Material combinations in MJ

One of the key features of MJ is its ability to create parts that include multiple materials at the same time in a single build. The materials are generally UV-curable photopolymers with a wide range of material properties, such as:

- Rigid and structural resins for strength
- Flexible and rubber-like materials for elasticity

- Transparent materials for visual inspection or optical use
- Colored resins for realistic prototypes and full-color printing

Material jetting systems can combine these materials by:

- Side-by-side deposition (e.g., rigid body with flexible joints)
- Gradient transitions, where properties change gradually across a part
- Digital material blending, where base resins are mixed at the droplet level to create custom composite materials with unique shore hardness, elasticity, or color.

1.2.1.2.3 Advantages of multi-material MJ

1. MJ can produce extremely fine layer heights (as low as 16 μm), which allows for detailed features, smooth surfaces, and precise geometry.
2. The ability to deposit multiple materials in one process makes it possible to manufacture complex assemblies with different physical and aesthetic properties.
3. MJ can use CMYK (cyan, magenta, yellow, and black) color mixing, making it possible to produce photo-realistic prototypes with color gradients, labels, and textures.
4. Parts often need little post-processing, and MJ is ideal for visual and presentation models.
5. Users can create hybrid materials with two or more base resins to achieve customized mechanical properties.
6. MJ is great for creating prototypes with real-world features such as soft- touch buttons, over-molded grips, and flexible joints, all in a single build cycle.
7. Since MJ uses UV curing and is a non-contact process, there is little material waste, and prints are clean and precise.

1.2.1.3 Selective laser sintering (SLS)

Selective laser sintering (SLS) is a common powder bed fusion (PBF) technology for additive manufacturing, which works by selectively fusing powdered material (generally polymer or metal) into 3D objects. While this was traditionally done with single-material fabrication, advances in SLS technology have allowed for what has been termed "multi-material" printing, or the capability to produce parts with different properties in the same 3D build (i.e., in a single print). This has opened the potential for new applications of SLS within the industry, such as aerospace, biomedical engineering, or electronics.

1.2.1.3.1 Working principle of SLS

SLS works by spreading a thin layer of powdered material (in certain cases, the part's design only takes place in specific areas of the powder bed while the rest remains untouched) onto the surface of the build platform. A form of laser beam scans the cross-section of the desired part; the laser will then heat to a certain extent, and sinter the powder particles together to form a solid layer. Once a layer is complete, the build platform lowers a set distance and compromises powder for the next layer. This printing process continues layer by layer until the 3D part is complete.

For SLS, there are multiple approaches to the multi-material principle.

- **Layer-by-Layer Material Deposition:** Different powders can be deposited into different areas of each layer.
- **Powder Mixing:** Powders can be pre-mixed together to make custom material blends that help to more locally control the material properties.
- **Voxel-Level Control:** Emerging systems are even allowing for material variation at the voxel level, which allows for each region in the layer to have different material compositions.

Upon completion of the print job, the part is removed from the unsintered powder bed, which serves as a natural support in the printing process, which is why SLS is excellent for printing complex geometries.

1.2.1.3.2 Material combinations in SLS

Multi-material PBF permits combining a broad range of powdered materials, including:

- Polymers with varying mechanical properties, including toughness, elasticity, and thermal resistance (e.g., PA12, TPU, and PEKK).
- Metal powders (in hybrid or experimental processes) fused with polymers for conductivity or structure.
- Ceramic-polymer composites for high-performance or bioactive applications.

1.2.1.3.3 Advantages of multi-material SLS

1. Parts with integrated mechanical, thermal, or electrical functions can be produced using multi-material SLS.
2. In crucial applications, seamless transitions between various material properties enhance performance and lessen stress concentrations.
3. Complex internal features and overhangs can be printed without the need for extra supports because the part is surrounded by unsintered powder during the printing process.
4. Because of their exceptional mechanical qualities, SLS parts are well-suited for practical, end-use applications.
5. Numerous polymers, composites, and even experimental metal-polymer mixtures can be employed.
6. Printing multipurpose components in one build reduces the need for post-print assembly, which saves money and time.

1.2.1.4 Vat photopolymerization (VP)

Vat photopolymerization (VP) is among the methods of 3D printing using the AM techniques, which produces 3D objects by selectively curing liquid photopolymer resin in a vat via light at certain wavelength. Examples of VP include digital light processing (DLP) and stereolithography (SLA). These technologies are widely acknowledged for their excellent surface finish, fine feature resolution, and precision, which makes them the perfect technologies for jewelers, microfluidics, dentists, and electronics. Thereby, complex components with spatially variable material properties can be generated in a single printing process due to new features that extended VP's capabilities for multi-material printing.

1.2.1.4.1 Working principle of VP

During the VP process, a vat containing liquid photopolymer resin is maintained. In accordance with the 3D model, portions of the resin are cured layer by layer with a light source, which can be a laser (for SLA) or a projector (for DLP). A fresh layer of resin is exposed and cured after a layer has dried, and the build platform either moves upward (in bottom-up systems) or downward (in top-down systems). Advanced mechanisms are integrated for printing multiple materials, including:

- Systems for exchanging resins, which enable the printer to alternate between various resins while building.
- Dispensing heads with multiple resins that can deposit different materials in specific areas.
- In order to modify the material's mechanical properties at the voxel level within a single resin, grayscale light exposure allows partial curing at varying intensities.

Parts with different stiffness, color, transparency, and thermal behavior can be produced using these methods.

1.2.1.4.2 Material combinations in VP

Multi-material vat photopolymerization utilizes various UV-curable resins, each tailored to offer specific properties:

- Rigid or flexible photopolymers for structural and elastic regions
- Transparent and colored resins for aesthetic and functional elements
- High-temperature or ceramic-filled resins for thermally resistant parts
- Biocompatible resins for medical and dental applications

These resins can be combined using dual/multi-vat systems or resin swapping to create the following:

- Functionally graded materials (FGMs)
- Multi-texture surfaces
- Embedded functional regions, such as flexible hinges within rigid bodies

1.2.1.4.3 Advantages of multi-material VP

1. For microscale applications, SLA and DLP technologies are perfect because they can achieve feature sizes as small as 25 μm.
2. VP parts are perfect for visual prototypes and final products because they usually require little post-processing.
3. VP improves part functionality by enabling gradual material blending or abrupt material transitions.
4. It reduces print time and contamination by allowing voxel-level control over material properties without changing physical resins.
5. It supports complex geometries, thin walls, and elaborate internal features.
6. VP is appropriate for implants, surgical guides, and medical models because certain resins are approved for direct human contact.

1.2.1.5 Direct ink writing (DIW)

Viscoelastic inks are used in direct ink writing (DIW), an extrusion-based additive manufacturing technique that creates 3D structures layer by layer. DIW is extremely versatile for a variety of materials because, in contrast to filament-based systems, it uses syringe-like nozzles to deposit material as pastes, gels, or slurries. Because DIW is a low-temperature, material-versatile process, it is especially well-suited for multi-material printing, which creates structures with integrated functionalities or graded properties by using several inks with different chemical or functional characteristics.

1.2.1.5.1 Working principle of DIW

In DIW, an ink with regulated rheological characteristics—that is, one that flows under pressure while maintaining its shape after deposition—is extruded onto a substrate via a tiny nozzle. A computer-controlled system that is based on a digital 3D model directs the print path. Several print heads or cartridges, each loaded with a distinct ink formulation, are used in multi-material DIW. To deposit spatially varied materials within a single structure, these inks can be switched out or used simultaneously.

Depending on the type of ink (e.g., polymer, ceramic, or metallic), post-processing techniques like drying, curing, or sintering may be used.

1.2.1.5.2 Material combinations in DIW

DIW is perfect for multi-material integration because it supports a large variety of structural and functional inks. Typical material kinds consist of:

- Conductive inks (e.g., silver nanoparticles, carbon-based inks) for embedded circuitry.
- Elastomers and soft polymers for soft robotics or flexible electronics.
- Biomaterials such as cell-laden hydrogels for tissue engineering.
- Ceramic slurries and metallic pastes for structural or thermal applications.

By combining different inks in a single build, DIW enables the fabrication of:

- Gradient mechanical properties, e.g., transitioning from soft to stiff zones.
- Multi-functional components, such as electronic devices with embedded conductors and insulators.
- Customizable biomimetic structures, replicating the complex makeup of natural tissues.

1.2.1.5.3 Advantages of multi-material DIW

1. DIW works with a wide variety of materials, such as biological, reactive, and viscous inks.
2. Because it runs at room temperature and low pressure, it can be used with materials that are sensitive to temperature, such as living cells.
3. DIW allows for the simultaneous integration of structure and function, as in the case of soft actuators with embedded sensing components.
4. It provides exact microscale control over architecture and material placement.
5. Low-cost implementation and versatility across multiple fields are made possible by a simple hardware setup.

Figure 1.1 illustrates the range of additive manufacturing processes adapted for multi-material capabilities.

Table 1.1 shows comparison between various additive manufacturing methods used for multi-material.

1.2.2 Multi-functional printing technologies

Multi-functional printing involves the integration of additional functionalities, such as sensors, actuators, or electronic circuits, within the printed parts. Several technologies have been developed to support multi-functional printing: Multi-functional printing is the act of adding additional functions, such as sensors, actuators, or electronic circuits, into printed parts. Multi-functional printing technologies have been discovered to perform these functions [4].

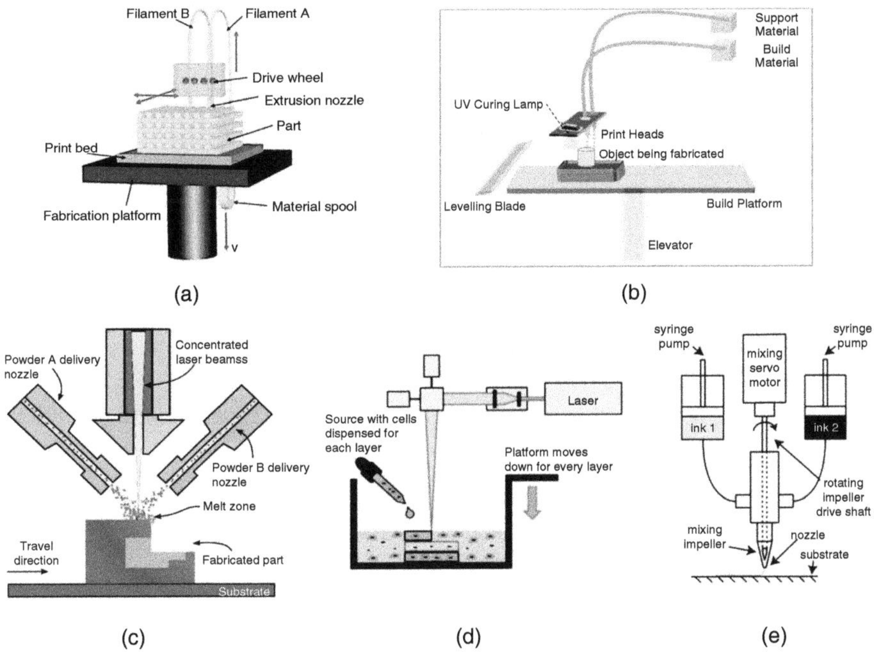

Figure 1.1 Schematic of a multi-material printing process using: (a) FDM, (b) material jetting, (c) selective laser sintering, (d) vat photopolymerization, and (e) direct ink writing.

Table 1.1 Comparison of multi-material printing technologies

Technology	*Materials supported*	*Resolution*	*Speed*	*Applications*
FDM	Thermoplastics	Medium	Medium	Prototyping, functional parts
Material jetting	Polymers, ceramics	Medium	High	Full-color models, functional parts
SLS	Polymers, metals	High	High	Aerospace, automotive
Vat photopolymerization	Photopolymers	High	Medium	Biomedical, optics
DIW	Polymers, ceramics, metals	High	Low	Electronics, biomedical

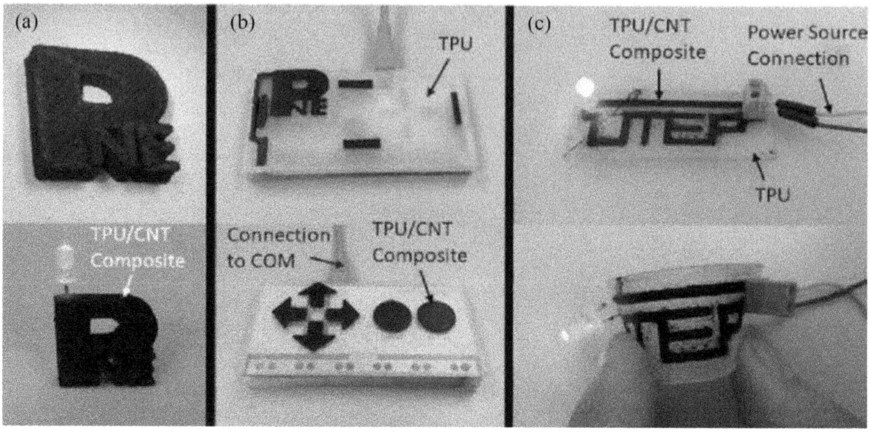

Figure 1.2 A 3D-printed object with integrated functionalities, such as conductive pathways (for electronics) and embedded sensors (for monitoring).

1.2.2.1 Embedded electronics

Multi-material printing can be used to integrate electronic devices, including conductive traces, resistors, and capacitors, into printed parts. This can be done by printing conductive inks or filaments along with insulating materials. Embedded electronics find widespread application in the manufacturing of smart devices, wearables, and Internet of Things (IoT) devices, as shown in Figure 1.2.

1.2.2.2 4D printing

4D printing is a new technology that uses smart materials that can change their shape or properties when exposed to some sort of stimulus, e.g. heat, light, or moisture. 4D multi-material printing offers the ability to create dynamic structures that can either build themselves or undergo changes based on the environment around them. It has the potential to be used in the construction of robotic systems, biomedical devices, and smart adaptive structures [5,6].

1.2.2.3 Hybrid manufacturing

Hybrid manufacturing integrates additive and subtractive manufacturing techniques into one machine. This enables the production of parts with topological diversity and integrated features. For instance, a hybrid manufacturing system may employ AM to extrude different materials and then employ CNC machining to produce accurate dimensions or finishes.

1.3 COMMON MATERIALS USED FOR MULTI-MATERIAL ADDITIVE MANUFACTURING

As a revolutionary technology in the field of manufacturing, multi-material AM has made it possible to integrate two or more materials with distinct properties in a single process, which could concatenate the formation of highly complex and multi-functional components. With this power, engineers and designers may design parts that are structurally and functionally tailored to specific uses. The performance requirements of the final product, for example, mechanical strength, flexibility, thermal conductivity, chemoresistance, and biological safety, determine the specific material selection in multi-material additive manufacturing [5].

1.3.1 Polymers

The wide range of materials that multi-material AM can process, beginning with polymers, which continue to be the most widely used because of their affordability, ease of printing, and widespread availability in both flexible and rigid forms, is a key enabler of the technology.

In fused deposition modeling (FDM) systems, thermoplastics like polyethylene terephthalate glycol (PETG), polylactic acid (PLA), and acrylonitrile butadiene styrene (ABS) are common polymers. ABS is valued for its toughness and impact resistance, PLA for its biodegradability and ease of use, and PETG for its strength and chemical resistance. Designers can produce parts with varying mechanical properties, surface finishes, and aesthetic characteristics all within a single print job with these materials.

But photopolymers used in MJ and SLA have more surface resolution, which allows producing the difficult and high-detailed parts. The wide spectrum of hardness, elasticity, and transparency that these light-sensitive resins can be adjusted, allowing for applications in dentistry, jewelry, and microfluidics.

Elastomers like silicone and thermoplastic polyurethane (TPU) are indispensable for stretchability, flexibility, or shock resistance in necessary applications. Elastomeric materials enhance user comfort and device conformity (they are used in, for instance, prosthetic liners, wearable health monitors, and soft robotics).

1.3.2 Metals

When it comes to high-performance applications that require structural integrity and resistance to corrosion, plastics face a formidable challenge with metals. Materials such as titanium, aluminum, and stainless steel are commonly processed in directed energy deposition (DED) and laser powder bed fusion (LPBF) machines.

It is a widely available and low- cost alloy with good corrosion resistance. Stainless steel is often used in applications that require superior corrosion resistance, high strength, or both.

With its exceptional strength-to-weight ratio and outstanding biocompatibility, titanium has become a metal of choice for a wide range of aerospace applications (turbine blades and frames of the aircraft) and medical needs (hip joint and cranial plates).

Aluminum is also lightweight and thermally conductive, making it a good choice for heat exchangers, engine enclosures, and structural components in electric cars and satellites. They can be mixed with polymers or ceramics in multi-material systems to build hybrid parts that leverage the strengths of each material. For example, metal-polymer components that integrate electrical shielding with impact resistance.

1.3.3 Ceramics

Among these materials, ceramics not only have superior properties in electrical insulation, biocompatibility, and high-temperature resistance, but are difficult to process as well. Components that are intended for use in severe conditions can be manufactured using AM processes that are specifically developed for ceramics, for instance, binder jetting and SLA with ceramic-containing slurries.

Because of its thermal and dielectric qualities, alumina (Al_2O_3) is frequently used in semiconductor manufacturing, engine insulation, and furnace components. Conversely, zirconia (ZrO_2) is highly valued for its strength, durability, and compatibility with human tissue, which makes it perfect for high-performance engineering ceramics, orthopedic devices, and dental implants.

1.3.4 Composites

Lastly, composite materials also help in breaking the mechanical strength barrier by combining the superior properties of various ingredients. Carbon fiber reinforced polymers (CFRP) are very popular in AM when it comes to printing robust and lightweight structures, and the field is well suited to applications such as drone or motorsport and aerospace (where lightweight is critical). Glass fiber reinforced polymers (GFRP) provide a less strong (marginally) but cheaper alternative, and are therefore suitable for elements in infrastructures, windmills, and cars. Together with other materials, these composites can be printed to produce parts with localized reinforcement, giving particular regions of a component precisely the right amount of strength and stiffness.

Advanced components with embedded electronics, multi-color or multi-texture surfaces, bio-inspired architectures, and functionally graded materials (FGMs) have been developed as a result of the integration of these various material classes within a single AM build. For example, it is now possible to print medical devices with embedded drug delivery channels, a soft- tissue interface, and a rigid structural frame. Similarly, parts for aerospace can have lightweight outer shells, heat-resistant cores, and conductive pathways without the need for post-processing or secondary assembly. Thus, industries like healthcare, aerospace, automotive, consumer electronics, and defense are changing as a result of the convergence of material science, software control, and advanced printing hardware. These innovations provide not only enhanced performance but also previously unheard-of design freedom.

1.4 APPLICATIONS OF MULTI-MATERIAL AND MULTI-FUNCTIONAL PRINTING

There are various applications in different fields for the multi-material multi-functional printing. Here are a few of the main use cases [7]:

1.4.1 Healthcare and biomedical

Multi-material and multi-functional printing will likely revolutionize healthcare and biomedical sectors by facilitating the development of personalized implants, prosthetic limbs, and medical devices. Top use cases include:

1.4.1.1 Customized implants

Multi-material printing enables the printing of patient-specific implants with built-in porous structures to promote bone ingrowth. As an example, the use of multi-material LPBF will enable the manufacture of titanium implants with integrated porous structures.

1.4.1.2 Prosthetics

Multi-material printing gives us the chance to use both hard and soft materials in one prosthetic device. This means we can make prosthetics that are strong, and also comfortable and easy to use. For example, a prosthetic arm might have a tough, hard-plastic frame along with soft, stretchy liners in areas that feel pressure. This makes it fit better and helps the user get less tired.

1.4.1.3 Bioprinting

Multi-material bioprinting lets us produce complex tissue constructs containing multiple cell types and extracellular matrices. This technology could be used in regenerative medicine and drug testing. Figure 1.3 shows the process of 3D bioprinting.

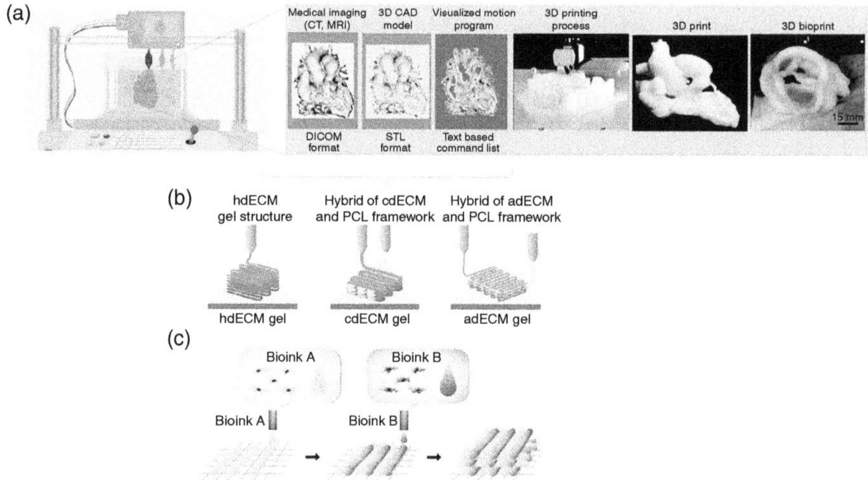

Figure 1.3 Process of 3D bioprinting, (a) steps of 3D bioprinting, (b) pre-scaffold fabrication bioprinting, (c) simultaneous hybrid 3D bioprinting.

1.4.2 Aerospace and automotive

The application of multi-material and multi-functional printing in aerospace and automotive fields has enabled lightweight and high-performance components with integrated functions. Key applications include the following.

1.4.2.1 Lightweight structures

Multi-material printing leads to the fabrication of lightweight structures with graded material properties. One use of this strong light metal is to mold aircraft parts made with titanium and aluminum.

1.4.2.2 Heat exchangers

This advanced multi-material printing method enables dynamic heat exchangers with cooling channels to be produced as single part. Multi-material LPBF, for example, can be used to produce aluminum heat exchangers with integrated cooling channels.

1.4.2.3 Embedded sensors

Printing multi-functional enables embedding of sensors in-situ into aerospace and automotive parts. These include the integration of strain gauges or temperature sensors into structural components to provide real-time performance monitoring.

1.4.3 Electronics and wearables

Multi-material and multi-functional printing is used in the electronics and wearables industry to create devices that are compact, lightweight, and functional. Among the top-line applications are:

1.4.3.1 Embedded electronics

Multi-material printing offers the integration of electronic components such as conductive traces, resistors, and capacitors directly into printed parts. The use of multi-material FDM or MJ can aid in the manufacturing of wearable devices that have integrated sensors and/or electronic circuits.

1.4.3.2 Flexible electronics

Printed electronics using multi-materials have recently emerged to create flexible and stretchable devices. For instance, teaching elastomeric materials and conductive inks allows for flexible circuits and sensors.

1.4.3.3 Smart textiles

Multi-material printing is utilized in smart textiles to integrate actuators, sensors, and electronic circuits. Thus, for example, wearables embedded with sensors to monitor vital signs can be fabricated using multi-material printing.

1.4.4 Consumer goods

The consumer goods sector has witnessed a revolution in creating customized and personalized products with the help of multi-material and multi-functional printing. The technology is used in a number of important areas:

1.4.4.1 Customized footwear

The production of personalized shoes with varying firmness and elasticity levels is now made possible using multi-material printing. As an example, multi-material FDM or MJ allows us to print running shoes with stiff soles and stretchy uppers.

1.4.4.2 Toys and games

Multi-material printing enables the creation of toys and games that integrate functions. For example, this technology enables the creation of toys with sensors or electronic circuits integrated inside them.

1.4.4.3 Home appliances

Home appliances with integrated functions are already possible due to multi-material printing. Kitchen appliances are one example where the method enables the use of sensors or heating elements in the product.

The emergence of multi-material and multi-functional printing has created new horizons in engineering, design, and customization in various industries. By enabling multiple materials and functionalities to coexist in one component, it not only improves performance and integration but also greatly simplifies production complexity and cost. With ongoing research and technological advancements, we can anticipate even more revolutionary applications in the near future.

1.5 CHALLENGES AND FUTURE DIRECTIONS

Although multi-functional and multi-material printing has taken great leaps, it still has some challenges. In order to get the best out of this technology, we must address these challenges. The following are the key issues and areas to investigate [8,9]:

1.5.1 Material compatibility

One of the largest challenges in printing with multiple materials is ensuring that they get along. When materials have varying melting points that expand with heat or have varying chemical compositions, it can be problematic. The layers may delaminate, the print may warp, or the materials may not adhere well to one another. We must discover materials that get along and adjust the printing process to ensure that various materials bond.

1.5.2 Process complexity

Printing with multiple materials and functions is no walk in the park. It needs precise control over how materials are laid down, hardened, and bonded. This makes both the machines and the computer programs that run them more complicated. We need to come up with better control systems and smarter ways to run the printing process to keep the quality consistent.

1.5.3 Cost and scalability

Printers that can handle multiple materials, and the materials themselves, can cost a pretty penny. This makes it hard for budget-conscious industries to jump on board. Also, it's tough to make more stuff without losing quality. We need to find ways to make materials and processes that don't break the bank, and figure out how to make more without sacrificing quality.

1.5.4 Post-processing

Multi-materials with the use of several materials that can be achieved by specific post-processing techniques to acquire the wanted surface or functionality. The process can increase the time and cost of the production phase. The further study should be on the production of these combined technologies together in a way that they can fit into the workflow of AM as seamless as possible.

1.5.5 Standardization and certification

The absence of industry standards and certification procedures for multi-material and multi-functional printing is a remarkable barrier to the industry, same as the aerospace and healthcare sectors. The future study should focus on the development of standardized testing and certification protocols, which would be used to ensure the quality of multi-material and multi-functional parts.

1.5.6 Integration with Industry 4.0

It is anticipated that the convergence of multi-material and multi-functional printing with Industry 4.0 devices, e.g., IoT, AI, and big data analytics, will bring more capabilities to AM. The upcoming study should concentrate on the creation of intelligent production systems which will employ these technologies intensively to get the design, manufacturing, and service delivery of multi-material and multi-functional parts.

1.6 CONCLUSION

Multi-material and multi-functional printing is a major step in additive manufacturing if we consider the fact that this form of manufacturing allows for the production of complex, high-performance parts that have functionalities included in them. This technology can be a game-changer in the medical, defense, electronics, and consumer goods industries, among others. There are some challenges to be tackled before a technology like the one we talked about (multi-material and multi-functional printing) can be leveraged "to the moon and beyond," beyond just that, these are material compatibility, process complexity, cost, scalability, post-processing, and standardization.

In the following, research should be focused on material systems that are compatible, control systems that are advanced, processes that are cost-effective, integrated post-processing solutions, and standardized testing protocols. Besides, the engagement of multi-material/multi-functional printing (MMP) and Industry 4.0 technologies can serve as a significant driver of AM and increase the attractiveness of various industries for innovation.

During the ongoing evolution of technology, multi-material and multi-functional printing will gain increasing recognition as a very important element of the manufacturing environment, and in such a way, we will witness a broad portfolio of products that will be relevant to clients in the digital era.

REFERENCES

1. Gibson, Ian, David Rosen, and Brent Stucker. 2015. *Additive Manufacturing Technologies: 3D Printing, Rapid Prototyping, and Direct Digital Manufacturing.* New York, NY: Springer.
2. Bandyopadhyay, Amit, and Susmita Bose. 2015. *Additive Manufacturing.* Boca Raton, FL: CRC Press.
3. Guo, Ning, and Ming C. Leu. 2013. "Additive Manufacturing: Technology, Applications and Research Needs." *Frontiers of Mechanical Engineering* 8 (3): 215–243.
4. Ngo, Tri D., Alireza Kashani, Giulio Imbalzano, Kate T. Q. Nguyen, and Dzung Hui. 2018. "Additive Manufacturing (3D Printing): A Review of Materials, Methods, Applications and Challenges." *Composites Part B: Engineering* 143: 172–196.
5. Momeni, Farnaz, Xiaoyue Liu, and Jun Ni. 2017. "A Review of 4D Printing." *Materials & Design* 122: 42–79.
6. Tibbits, Skylar. 2014. "4D Printing: Multi-Material Shape Change." *Architectural Design* 84 (1): 116–121.
7. Zhang, Yiheng, and Yuwei Zhang. 2019. "Multi-Material Additive Manufacturing: A Systematic Review." *Journal of Manufacturing Processes* 45: 1–16.
8. Gardan, Julien. 2016. "Smart Materials in Additive Manufacturing: State of the Art and Perspectives." *Materials* 9 (10): 787.
9. Yap, Yau L., Wai Y. Yeong, and Hong K. Tan. 2015. "Multimaterial 3D Printing of Soft and Rigid Components for Functional Robotic Devices." *Advanced Materials Technologies* 1 (4): 1600045.

Chapter 2

Recent advances in additive manufacturing through AI applications

Sachin Sirohi, Upender Punia, Ankit Tyagi, Rekha Nair, Ankita Dadhich, and Abhishek Dadhich

2.1 INTRODUCTION

Metal additive manufacturing (AM), sometimes known as metal 3D printing, has evolved from a simple method of producing prototypes to a fully formed manufacturing process that challenges conventional wisdom [1–3]. Metal AM has allowed for shorter production times, less waste, and greater design freedom since the 1990s, allowing for the creation of complex, high-performance parts. Geometric complexity, lightweighting, and functional integration are extensively employed in the energy, healthcare, automotive, and aviation sectors as a result of their practicality [4–6]. Metal AM, in contrast to subtractive or formative manufacturing, employs digital models to develop components layer by layer. This facilitates the development of lattice structures, intricate internal channels, and topology-optimised forms [3,7,8]. The efficacy and capacity of a building to perform multiple functions may be enhanced by the incorporation of numerous functionally graded materials into a single structure in buildings constructed with AM [9,10]. These qualities are the reason why metal additive manufacturing is essential for Industry 4.0 and smart output [11,12]. PBF, DED, and BJ are the metal AM techniques that are employed most frequently in both academic and professional environments [5,13,14]. Electron beam melting (EBM) and selective laser melting (SLM) are two widely used techniques for the production of microstructurally controlled components that are precise [15,16]. By depositing material onto surfaces, DED technologies enable large-scale production, repair, and multi-material fabrication [17,18]. After sintering or infiltration, binder jetting is an excellent technique for mass manufacturing due to its high productivity and minimal shape change during heating [14,19]. Additive manufacturing in metals does have its uses, but it also has drawbacks that make it unique among manufacturing techniques. AM's rapid heating and cooling cycles create temperature gradients, non-equilibrium solidification conditions, and complex thermal histories, which affect microstructural development and mechanical behaviour [15,16,20]. These circumstances cause anisotropic characteristics, residual stresses, distortion, porosity, lack of fusion, keyholing, and

DOI: 10.1201/9781003607052-2

solidification cracking [21–24]. Thus, AM-treated casting or wrought alloys perform poorly [25,26]. This realisation has led to AM-specific materials design, where alloys are developed for service performance, printability, process stability, and fault tolerance [16,25,27]. Crack-resistant aluminium alloys are made by altering solidification and adding grain-refining components to prevent hot tearing [25,26]. Customised nickel-based superalloys and stainless steels balance printability, high-temperature strength, and corrosion resistance [28,29]. High-entropy alloys (HEAs) in AM support this trend by offering customisable compositions, improved thermal stability, and higher strength-ductility [30–32]. Metal AM build quality and repeatability depend on powder feedstock factors. Particle size distribution, morphology, flowability, surface chemistry, and oxygen concentration greatly affect powder spreading, melt pool stability, and defect development [33–36]. Gas atomisation, plasma spheroidisation, and powder conditioning have enhanced feedstock homogeneity and recyclability, making processes more resilient and sustainable [34–37]. Powder deterioration during reuse remains a problem, spurring research into powder life-cycle management and in-situ quality monitoring [38,39]. AI has disrupted metal AM alongside material advancements. Trial-and-error optimisation is inefficient and expensive because of the AM parameter space complexity and nonlinear interactions between process variables, microstructure, and features [16,27,40]. Machine learning, deep learning, and reinforcement learning (RL) provide data-driven modelling, prediction, and optimisation throughout the AM workflow, making them tempting options [40–43]. Machine learning in metal AM first predicted porosity, surface roughness, and mechanical parameters from process data using regression models and artificial neural networks [41,44]. Recent research has employed CNNs and RNNs to analyse high-resolution in-situ sensor data such as melt pool pictures, thermal signatures, and acoustic emissions for real-time defect diagnosis and process monitoring [45–47]. These methods accurately detect flaws such as absence of fusion, spatter development, and keyhole instability, enabling proactive quality assurance and reduced scrap rates [45,48,49].

Using AI with physics-based models improves prediction and interpretation. Hybrid frameworks that use finite element simulations, computational fluid dynamics, thermodynamic modelling, and machine learning surrogates may rapidly explore large parameter spaces while retaining physical consistency [16,27,50]. Digital twins, virtual copies of AM systems, provide simulation-driven process optimisation, predictive maintenance, and closed-loop control [51–53]. Digital twins underpin intelligent industrial ecosystems [11,51]. AI has also aided AM materials research and design. Data-driven alloy composition screening using high-throughput simulations, materials databases, and machine learning algorithms can quickly assess printability, strength, corrosion resistance, and thermal stability [54–57]. ML-assisted frameworks have identified crack-resistant aluminium alloys, improved AM steel heat treatment, and predicted phase stability in HEAs

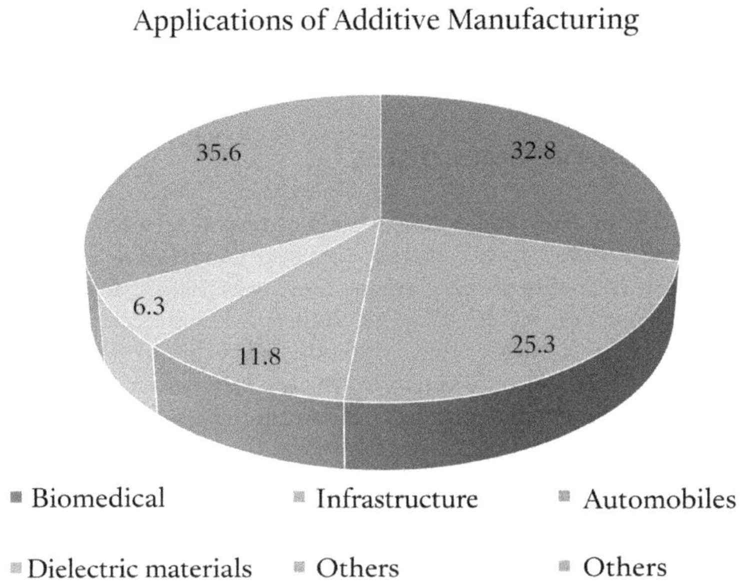

Figure 2.1 Applications of additive manufacturing.

under non-equilibrium solidification [25,26,32]. These methods reduce experimental costs and time, speeding alloy idea to industrial approval [55,58]. AI supports sophisticated design for additive manufacturing (DfAM) through generative design, topology optimisation, and multi-objective trade-off analysis that considers manufacturability, performance, cost, and sustainability [59–62]. Optimising building orientation, support structures, and scanning methods with reinforcement learning and evolutionary algorithms improves surface quality, residual stress, and post-processing [62–64]. Since they optimise material efficiency, energy savings, and lifespan performance, intelligent design strategies support sustainable manufacturing and the circular economy [65–67]. Despite significant advances, AI-enabled metal AM cannot be fully industrialised. Since private constraints, limited standardisation, and machine and material heterogeneity make generalised and transferable models challenging, data availability and quality remain major issues [27,40,45]. Black-box AI models' interpretability complicates safety-critical applications, certification, and regulatory compliance, especially in biomedical and aerospace [51–53,68–71].

Lastly, metal additive manufacturing (AM) is a massive integration of AI and materials science that is revolutionising production and application in industry as shown in Figure 2.1. Users have unprecedented control over the process's design, features, and performance because of the combination of AI-driven process intelligence and the creation of AM-specific materials. The process is also more efficient, dependable, and long-lasting. The next generation of smart, customisable, and long-lasting manufacturing systems will include metal AM when data-driven methodologies and fundamental physical knowledge are combined [16,27,51,53,71].

2.2 METAL AM TECHNOLOGIES

Using 3D CAD models, metal AM manufactures metal components layer-by- layer [1–3]. Input types, energy sources, temperature histories, and post-processing requirements vary among these approaches, impacting microstructure and function [4,13,72,73]. See Tables 2.1 and 2.2 for a variety of metal additive manufacturing types. Methods such as binder jetting (BJ), powder bed fusion (PBF), sheet lamination, material extrusion, directed energy deposition (DED), and many more are included in this category [4,13,14]. Each approach presents unique heat management, residual stress production, defect development, and microstructural control challenges, requiring advanced materials selection, process optimisation, and quality assurance [4,14,27].

Table 2.1 Comparison of major metal AM technologies

Technology	*Energy source*	*Feedstock form*	*Typical applications*	*Key advantages*	*Key limitations*
Powder bed fusion (PBF)	Laser/ electron beam	Powder	Aerospace parts, medical implants, tooling	High precision, fine microstructure, complex geometry	Residual stresses, porosity, high equipment cost
Directed energy deposition (DED)	Laser/ electron beam/ plasma	Powder or wire	Repair, large structures, functionally graded materials	High deposition rate, multi-material capability, large build size	Coarser microstructure, lower surface finish, dimensional inaccuracy
Binder jetting (BJ)	No melting during printing	Powder+ binder	Mass production, complex shapes, tooling	Fast build speed, low thermal distortion, scalable	Requires sintering, shrinkage control, lower density
Sheet lamination	Ultrasonic/ adhesive/ diffusion bonding	Metal sheets	Embedded electronics, layered structures	Low thermal distortion, high build rate	Limited materials, weaker interlayer bonding
Material extrusion	Mechanical extrusion+ furnace	Metal-polymer filament or paste	Prototyping, low-cost manufacturing	Low equipment cost, easy operation	Lower density, dimensional inaccuracy, limited mechanical performance

Table 2.2 Thermal, microstructural, and defect characteristics

Technology	*Thermal gradients*	*Microstructure*	*Common defects*	*Post-processing needs*
PBF	Very high	Fine, non-equilibrium, columnar grains	Porosity, lack of fusion, cracking	Stress relief, HIP, machining
DED	Moderate to high	Coarser grains, heterogeneous	Dilution, residual stress, distortion	Machining, heat treatment
BJ	Low during printing; high during sintering	Powder-sintered, isotropic (after sintering)	Shrinkage, porosity, distortion	Debinding, sintering, infiltration
Sheet lamination	Low	Layered, anisotropic bonding	Delamination, weak interfaces	Machining, bonding enhancement
Material extrusion	Low during printing; high during sintering	Powder-sintered	Shrinkage, porosity, distortion	Debinding, sintering, machining

2.2.1 Powder bed fusion (PBF)

High-precision, surface polishing, and microstructural control industries use PBF metal additive fabrication [4,13,15]. PBF uses a laser or electron beam to selectively melt or sinter computer-designed regions on a metal powder-coated build platform. PBF produces complicated geometries, internal cooling channels, lattice structures, and lightweight components that conventional manufacturing cannot achieve [2,3,7,8]. PBF's fast heating and cooling cycles cause steep thermal gradients and high solidification rates, resulting in non-equilibrium microstructures, residual stresses, and anisotropic mechanical characteristics [15,16,20]. Warping, deformation, and cracking can occur in alloys with a narrow solidification range or high hot tearing sensitivity [21–24,25]. A lack of fusion and porosity can cause structural weakness in melt pool instability, spatter development, and powder bed defects [21–23,33]. Thus, extensive research has optimised scan techniques, layer thickness, energy density, shielding settings, and AM-specific alloys for PBF weldability, fracture resistance, and microstructural stability [16,25–29]. For industrial performance, hot isostatic pressing (HIP), stress relief annealing, and surface polishing are often needed [5,16,21].

2.2.2 Directed energy deposition (DED)

DED uses a laser, electron beam, or plasma arc to melt metal wire or powder and deposit it on a substrate or pre-existing component [17,18]. DED systems can create massive structures, repair portions, and add functionality to pre-existing components in an open environment, unlike PBF [17,18]. DED has high deposition speeds, scalability to large geometries, and the ability to dynamically change feedstock composition during deposition to create functionally graded materials [9,17,18]. These properties make DED ideal for aerospace, defence, and heavy industrial applications such as turbine blade maintenance, tooling refurbishing, and structural strengthening [5,18,72]. However, DED's open nature and higher energy input cause bigger melt pools and slower cooling rates than PBF, resulting in coarser microstructures and mechanical variability [17,20,28]. Complex temperature cycles and multi-layer reheating can increase residual stress, interface dilution, and compositional variability in multi-material systems [17,20,27]. Maintaining dimensional precision and surface smoothness in DED requires considerable post-processing via machining or surface treatment [17,18,23]. Advanced process monitoring, closed-loop control, and AI-driven parameter optimisation are used to stabilise melt pool behaviour, eliminate defects, and increase DED component microstructural consistency [40–47,50].

2.2.3 Binder jetting (BJ)

Unlike metal AM, BJ separates form formation and densification [14,19]. Selectively applying a liquid binder over a powder bed bonds metal powder particles to generate a "green" portion in BJ [14,19]. Thermal post-processing debinds, sinters, and sometimes infiltrates the green component to densify and strengthen it [14,19]. BJ allows for quicker build rates, a wide variety of alloys, and reduced thermal distortion during printing, as no melting occurs during component creation [14,19]. BJ excels in high-volume manufacturing, complicated geometries, and cost-effective and scalable applications [14,19]. However, BJ causes dimensional shrinkage, porosity control, and mechanical property variability during sintering [14,19,33,36]. Powder properties, binder chemistry, debinding profiles, and sintering temperatures must be controlled to achieve near-full density and dimensional accuracy [14,19,33,36]. For structural applications, process integration and validation are essential since post-processing factors strongly influence BJ component mechanical characteristics [5,14]. BJ research aims to increase powder packing density, design binders with cleaner burnout, and employ AI-driven sintering models to accurately predict shrinkage, densification, and microstructural evolution [40,50,54–58].

2.2.4 Sheet lamination and material extrusion

For some applications and alloy systems, sheet lamination and material extrusion are better metal AM methods [2,59]. Sheet lamination comprises ultrasonic welding, diffusion bonding, or adhesive bonding thin metal sheets or foils, and selectively cutting to shape [2,59]. It offers reduced thermal distortion, quicker assembly, and the option to add integrated components or sensors during production [2,59]. Sheet lamination is restricted by the availability of appropriate sheet materials, the difficulties of strong interlayer bonding, and the complexity of manufacturing fine internal features compared to powder-based AM [2,59]. Therefore, its use is limited to specialised industrial applications [2,59]. Extruded metal-filled polymer filaments or pastes are debound and sintered like BJ [14,19,59]. This method is ideal for prototyping, teaching, and small-batch manufacturing due to its low cost, ease of use, and compatibility with office-scale 3D printers [14,59]. Component density, dimensional precision, and mechanical performance restrict its applicability in high-performance structural applications [5,14,19].

2.2.5 Cross-cutting challenges and opportunities

All metal AM methods struggle with temperature management, defect control, microstructural heterogeneity, and process reproducibility [4,16,20,27]. Residual stresses, phase transformations, and texture development from steep temperature gradients and frequent reheating impact mechanical performance, fatigue life, and corrosion resistance [16,20,21,28]. Maintaining component quality is difficult due to porosity, lack of fusion, inclusions, and surface roughness [21–23,33,36]. We need physics-based modelling, real-time process monitoring, improved materials design, and AI-driven optimisation to tackle these problems [16,27,40,50]. AM-specific alloys, feedstock management, and adaptive process control systems are improving production reliability [25–29,33–37].

2.3 ARTIFICIAL INTELLIGENCE IN METAL AM

The complexity of metal AM processes is on the rise, as there are nonlinear connections between process parameters, thermal fields, microstructural development, and mechanical performance [4,15,72,74]. Consequently, conventional optimisation strategies that are predicated on trial- and -error are both costly and ineffective. Data analytics, machine learning, deep learning, and reinforcement learning have revolutionised data-driven metal AM systems that are smart and self-driving [11,75–77]. By gaining a deeper understanding of the process, generating predictions, keeping tabs in real time, implementing adaptive control, and discovering novel materials, AI enhances the quality, reliability, and manufacturing speed of components [41,78–80]. Laser power, scan speed, hatch spacing, layer thickness, powder characteristics, and atmosphere are among the many factors that influence metal AM [81–83]. These parameters alter the microstructure, mechanical characteristics, melt pool behaviour, defect development, and nonlinearity in complex ways [84–86]. While helpful, classical physics models are computationally intensive and slow to explain events occurring on several scales [87–89]. Quickly predicting the outcome of a process is possible with the use of AI-powered surrogate models trained on simulation and trial data [90–92]. Predictions of porosity, surface roughness, dimensional accuracy, and mechanical performance have been made using regression models, ANNs, SVMs, and ensemble learning due to process variables [93–95]. The optimal parameter windows with the least amount of testing may be found using these models, which provide tools for multi-objective planning that look at quality, productivity, and cost [96–98]. With RL, self-aware agents may optimise processes by investigating and adjusting parameters in response to interactions in the production environment [99–101]. Controls based on RL may dynamically adjust the laser power, scan speed, and wire feed rate to maintain a steady melt pool and good build quality in various scenarios [102–105].

2.4 CONCLUSIONS

The energy, aviation, medical, and automobile sectors have all benefited from the new ways that metal additive manufacturing simplifies the design, production, and usage of complex, high-performance metal components. Additive manufacturing techniques such as binder jetting, material extrusion, powder bed fusion, directed energy deposition, and sheet lamination have simplified product development but increased the difficulty of temperature control, structural stability, defect elimination, and process reliability. Modern material science has resulted in feedstock systems and metals that are optimal for additive processing. With them, issues with solidification cracking, anisotropy, porosity, and wear performance have been long-lived. Thanks to meticulous metal design, optimisation of post-processing, and microstructure engineering, metal AM has progressed from prototype to production ready for approval and end-of-life. Artificial intelligence (AI) has the potential to improve this scenario through its fault detection, real-time tracking, predictive modelling, digital twin generation, and closed-loop process management capabilities. DfAM, material discovery, and process-structure-property correlations are all enhanced by AI-powered systems. This reduces waste, shortens manufacturing timelines, and enhances component quality. Autonomous, self-improving AM systems that can learn and make decisions autonomously are enabled by sensor-rich industrial environments and artificial intelligence (AI). Despite these significant advancements, the issues of data availability, model generalisability, interpretability, standardisation, and regulation approval have not been entirely resolved, particularly for safety-critical applications. In order to surmount these obstacles, it would be imperative to engage in collaborative endeavours regarding data infrastructure, physically grounded machine learning, explicable artificial intelligence, and cross-platform assessment tools. The incorporation of modern metal additive printing technology, smart material design, and AI will result in manufacturing systems that are more durable, environmentally friendly, and intelligent. Metal AM is a technology that holds significant potential for sustainable industrial development and next-generation engineering. However, complete realisation necessitates cross-disciplinary research and practical application in the business sector.

REFERENCES

1. Kruth, J.P., Leu, M.C., & Nakagawa, T. (1998). Progress in AM and rapid prototyping. *CIRP Annals*, 47(2), 525–540.
2. Gibson, I., Rosen, D., & Stucker, B. (2015). *Additive Manufacturing Technologies: 3D Printing, Rapid Prototyping, and Direct Digital Manufacturing* (2nd ed.). New York, NY: Springer.
3. Thompson, M.K., et al. (2016). Design for AM: Trends, opportunities, considerations, and constraints. *CIRP Annals*, 65(2), 737–760.
4. DebRoy, T., et al. (2018). AM of metallic components – Process, structure and properties. *Progress in Materials Science*, 92, 112–224.
5. Lewandowski, J.J., & Seifi, M. (2016). Metal AM: A review of mechanical properties. *Annual Review of Materials Research*, 46, 151–186.
6. Gu, D., Meiners, W., Wissenbach, K., & Poprawe, R. (2012). Laser AM of metallic components. *International Materials Reviews*, 57(3), 133–164.
7. Rosen, D.W. (2014). Design for AM: A method to explore unexplored regions of the design space. *Proceedings of the ASME 2014 International Design Engineering Technical Conferences and Computers and Information in Engineering Conference (IDETC/CIE 2014)*, August 17–20, 2014, Buffalo, New York, USA. New York, NY: American Society of Mechanical Engineers (ASME).
8. Leary, M., Merli, L., Torti, F., Mazur, M., & Brandt, M. (2014). Optimal topology for additive manufacture. *Materials & Design*, 63, 678–690.
9. Bandyopadhyay, A., & Heer, B. (2018). AM of multi-material structures. *Materials Science and Engineering: R*, 129, 1–16.
10. Zhang, Y., et al. (2018). Functionally graded materials in AM. *Composites Part B*, 153, 336–347.
11. Kusiak, A. (2018). Smart manufacturing. *International Journal of Production Research*, 56(1–2), 508–517.
12. Lu, Y., Liu, C., Wang, I., Huang, H., & Xu, X. (2021). Digital twin-driven smart manufacturing. *Journal of Manufacturing Systems*, 58, 346–361.
13. Herzog, D., Seyda, V., Wycisk, E., & Emmelmann, C. (2016). AM of metals. *Acta Materialia*, 117, 371–392.
14. Mostafaei, A., et al. (2021). Binder jetting of metallic materials. *Progress in Materials Science*, 119, 100707.
15. King, W.E., et al. (2015). Laser powder bed fusion AM of metals; physics, computational, and materials challenges. *Applied Physics Reviews*, 2, 041304.
16. DebRoy, T., et al. (2019). Scientific challenges in metal AM. *Science*, 364(6435), eaav 483.
17. Pinkerton, A.J. (2016). Advances in the modeling of laser direct metal deposition. *Journal of Laser Applications*, 27(S2), S15001.
18. Beaman, J.J., Bourell, D.L., Seepersad, C.C., & Kovar, D. (2020). *AM Review*. U.S. Department of Energy.

19. ExOne. (2020). *Binder Jetting Technology Overview*. ExOne Technical White Paper.
20. Qiu, C., Adkins, N.J.E., & Attallah, M.M. (2015). Microstructure and mechanical properties of AlSi10Mg produced by selective laser melting. *Materials Science and Engineering A*, 578, 230–239.
21. Gong, H., Rafi, K., Gu, H., Janaki Ram, G.D., Starr, T., & Stucker, B. (2015). Analysis of defect generation in Ti–6Al–4V parts made using SLM. *AM*, 1–4, 87–98.
22. Cunningham, R., Narra, S.P., Montgomery, C., Beuth, J., & Rollett, A.D. (2017). Synchrotron-based X-ray microtomography characterization of AM materials. *JOM*, 69, 479–484.
23. Gockel, J., Sheridan, L., Koerper, B., & Whip, B. (2014). The influence of AM processing parameters on surface roughness and fatigue performance. *Materials & Design*, 65, 739–748.
24. Carter, L.N., Attallah, M.M., & Reed, R.C. (2012). Laser powder bed fabrication of nickel-base superalloys. *Materials Science and Engineering A*, 536, 136–144.
25. Martin, J.H., et al. (2017). 3D printing of high-strength aluminum alloys. *Nature*, 549, 365–369.
26. Collins, P.C., Brice, D.A., Samimi, P., Ghamarian, I., & Fraser, H.L. (2016). Microstructural control of additively manufactured metallic materials. *Annual Review of Materials Research*, 46, 63–91.
27. DebRoy, T., et al. (2019). AM of metallic components: The science. *Science*, 364, eaav483.
28. Zhang, B., et al. (2019). AM of nickel-based superalloys. *Journal of Materials Science & Technology*, 35, 201–212.
29. Li, X., et al. (2020). Stainless steels fabricated by laser powder bed fusion. *Materials Science and Engineering A*, 785, 139395.
30. Cantor, B., Chang, I.T.H., Knight, P., & Vincent, A.J.B. (2004). Microstructural development in equiatomic multicomponent alloys. *Materials Science and Engineering A*, 375–377, 213–218.
31. Miracle, D.B., & Senkov, O.N. (2017). A critical review of high entropy alloys. *Acta Materialia*, 122, 448–511.
32. Yang, T., et al. (2020). AM of high entropy alloys. *Materials Today*, 30, 67–83.
33. Slotwinski, J.A., Garboczi, E.J., & Hebenstreit, K.M. (2014). Porosity measurements and analysis for metal AM parts. *Journal of Research of NIST*, 119, 494–528.
34. Spierings, A.B., Levy, G., Wegener, K. (2012). Powder flowability characterization methodology for AM. *Rapid Prototyping Journal*, 18(5), 409–417.
35. Leung, C.L.A., et al. (2018). In situ X-ray imaging of melt pool dynamics. *Acta Materialia*, 166, 294–305.
36. Sutton, A.T., et al. (2020). Powder reuse effects in laser powder bed fusion. *AM*, 36, 101439.
37. Clayton, P., et al. (2017). Powder recycling and its effect on metal AM. *Metallurgical and Materials Transactions A*, 48, 3773–3782.
38. Tang, M., et al. (2015). Defects in powder reuse for AM. *Materials Science and Engineering A*, 622, 41–46.

39. Slotwinski, J.A., et al. (2015). *Metal Powder Life Cycle in AM*. NIST Report.
40. Bock, F.E., et al. (2019). A review of machine learning in materials science. *NPJ Computational Materials*, 5, 84.
41. Tapia, G., & Elwany, A. (2014). A review on process monitoring and control in metal-based AM. *Journal of Manufacturing Science and Engineering*, 136(6), 060801.
42. Liu, Z., et al. (2020). Machine learning in AM. *Journal of Manufacturing Processes*, 57, 654–669.
43. Scime, L., & Beuth, J. (2018). A multi-scale convolutional neural network for defect classification in AM. *AM*, 24, 273–286.
44. Garg, A., et al. (2018). Machine learning approaches for AM process optimization. *Journal of Manufacturing Processes*, 32, 626–636.
45. Everton, S.K., et al. (2016). Review of in-situ process monitoring and in-process metrology for metal AM. *Materials & Design*, 95, 431–445.
46. Zhang, Y., et al. (2020). Deep learning for melt pool defect detection. *AM*, 36, 101447.
47. Ye, D., et al. (2018). In situ monitoring and data-driven analysis in AM. *Journal of Manufacturing Processes*, 36, 155–168.
48. Craeghs, T., et al. (2011). Online quality control of SLM. *Rapid Prototyping Journal*, 17(5), 313–320.
49. Lane, B., et al. (2016). Thermographic monitoring of metal AM. NIST Technical Note.
50. Gan, Z., et al. (2019). Data-driven modeling for AM microstructure prediction. *Computational Materials Science*, 163, 239–246.
51. Tao, F., et al. (2018). Digital twin in industry: State-of-the-art. *IEEE Transactions on Industrial Informatics*, 15(4), 2405–2415.
52. Zhang, H., et al. (2021). Digital twins for AM. *Journal of Manufacturing Systems*, 61, 1–15.
53. Lu, Y., et al. (2021). Digital twin-driven smart manufacturing: Connotation, reference model, and applications. *Robotics and Computer-Integrated Manufacturing*, 61, 101837.
54. Agrawal, A., & Choudhary, A. (2016). Perspective: Materials informatics and big data. *APL Materials*, 4, 053208.
55. Butler, K.T., Davies, D.W., Cartwright, H., Isayev, O., & Walsh, A. (2018). Machine learning for molecular and materials science. *Nature*, 559, 547–555.
56. Rajan, K. (2015). Materials informatics. *Materials Today*, 18(8), 471–478.
57. Kalidindi, S.R., & De Graef, M. (2015). Materials data science. *Annual Review of Materials Research*, 45, 171–193.
58. Lookman, T., et al. (2019). Active learning in materials science. *NPJ Computational Materials*, 5, 21.
59. Gao, W., et al. (2015). The status, challenges, and future of AM. *Computer-Aided Design*, 69, 65–89.
60. Wu, J., Wang, X., Wang, L., & Ding, H. (2016). Topology optimization for AM. *Structural and Multidisciplinary Optimization*, 54, 935–948.
61. Chen, Y., et al. (2020). Machine learning-based design for AM. *Engineering*, 6(5), 536–547.
62. Rosen, D.W. (2014). Design for AM. *CIRP Annals*, 63(2), 535–552.

63. Wu, J., et al. (2016). Build orientation optimization using machine learning. *Journal of Manufacturing Processes*, 24, 101–110.
64. Leary, M., et al. (2014). Design for AM: A critical review. *Materials & Design*, 63, 678–690.
65. Ford, S., & Despeisse, M. (2016). AM and sustainability. *Journal of Cleaner Production*, 137, 1573–1587.
66. Kellens, K., et al. (2017). Environmental impact of AM. *Journal of Industrial Ecology*, 21, S49–S68.
67. Gebler, M., Schoot Uiterkamp, A.J.M., & Visser, C. (2014). A global sustainability perspective on 3D printing. *Energy Policy*, 74, 158–167.
68. Seifi, M., et al. (2016). Progress towards metal AM standardization. *JOM*, 68, 747–764.
69. Ribeiro, M.T., Singh, S., & Guestrin, C. (2016). Why should I trust you? Explaining the predictions of any classifier. *Proceedings of the 22nd ACM SIGKDD International Conference on Knowledge Discovery and Data Mining (KDD '16)*, August 13–17, 2016, San Francisco, California, USA. New York, NY: Association for Computing Machinery (ACM), pp. 1135–1144.
70. Karniadakis, G.E., et al. (2021). Physics-informed machine learning. *Nature Reviews Physics*, 3, 422–440.
71. Willard, J., et al. (2022). Integrating physics-based modeling with machine learning. *Nature Reviews Physics*, 4, 789–807.
72. Frazier, W.E. (2014). Metal AM: A review. *Journal of Materials Engineering and Performance*, 23(6), 1917–1928.
73. Wohlers, T., & Gornet, T. (2014). *History of AM*. Wohlers Report.
74. Gibson, I., Rosen, D.W., & Stucker, B. (2021). *Additive Manufacturing Technologies* (3rd ed.). Cham: Springer.
75. Tao, F., Qi, Q., Liu, A., & Kusiak, A. (2018). Data-driven smart manufacturing. *Journal of Manufacturing Systems*, 48, 157–169.
76. Bocklisch, T. (2016). Hybrid energy storage approach for renewable energy applications. *Energy Procedia*, 105, 153–158.
77. Wuest, T., Weimer, D., Irgens, C., & Thoben, K.D. (2016). Machine learning in manufacturing: Advantages, challenges, and applications. *Production & Manufacturing Research*, 4(1), 23–45.
78. Grasso, M., & Colosimo, B.M. (2017). Process defects and in situ monitoring methods in metal powder bed fusion. *Measurement Science and Technology*, 28(4), 044005.
79. Everton, S.K., et al. (2016). Review of in-situ process monitoring and in-situ metrology for metal AM. *Materials & Design*, 95, 431–445.
80. Leal, R., et al. (2017). AM tooling for the automotive industry. *The International Journal of Advanced Manufacturing Technology*, 92, 1671–1676.
81. Craeghs, T., et al. (2012). Online quality control of selective laser melting. *Proceedings of the Solid Freeform Fabrication Symposium (SFF Symposium 2012)*, August 6–8, 2012, Austin, Texas, USA. Austin, TX: The University of Texas at Austin, pp. 212–226.
82. Zhang, Y., Bernard, A., Harik, R., & Karunakaran, K.P. (2017). Build orientation optimization for AM. *Journal of Manufacturing Systems*, 45, 133–145.
83. Yadroitsev, I., & Smurov, I. (2011). Selective laser melting technology. *Physics Procedia*, 12, 264–270.

84. King, W.E., et al. (2014). Observation of keyhole-mode laser melting in laser powder-bed fusion. *Journal of Materials Processing Technology*, 214(12), 2915–2925.
85. Thijs, L., et al. (2010). A study of the microstructural evolution during selective laser melting of Ti–6Al–4V. *Acta Materialia*, 58(9), 3303–3312.
86. Gong, H., et al. (2015). Influence of defects on mechanical properties of Ti-6Al-4V components produced by selective laser melting. *Materials & Design*, 86, 545–554.
87. Khairallah, S.A., Anderson, A.T., Rubenchik, A., & King, W.E. (2016). Laser powder-bed fusion AM: Physics of complex melt flow and formation mechanisms of pores, spatter, and denudation zones. *Acta Materialia*, 108, 36–45.
88. Roberts, I.A., et al. (2009). A three-dimensional finite element analysis of the temperature field during laser melting of metal powders. *International Journal of Machine Tools and Manufacture*, 49(12–13), 916–923.
89. Francois, M.M., et al. (2017). Modeling of AM processes for metals: Challenges and opportunities. *Current Opinion in Solid State and Materials Science*, 21(4), 198–206.
90. Zhang, L., et al. (2019). Machine learning modeling of selective laser melting process parameters and part quality. *Journal of Manufacturing Processes*, 45, 444–453.
91. Tapia, G., et al. (2016). Modeling and control of melt pool dynamics in metal AM. *Journal of Manufacturing Science and Engineering*, 138(9), 091006.
92. Scime, L., & Beuth, J. (2019). Anomaly detection and classification in a laser powder bed AM process using a trained computer vision algorithm. *AM*, 19, 114–126.
93. Qi, H., Azer, M., & Ritter, A. (2010). Studies of standard heat treatment effects on microstructure and mechanical properties of laser net shape manufactured Inconel 718. *Metallurgical and Materials Transactions A*, 41(8), 2008–2022.
94. Caiazzo, F., Alfieri, V., Casalino, G., & Guarino, S. (2013). AM by means of laser-aided directed metal deposition. *Materials & Design*, 51, 834–842.
95. Balu, P., Leggett, P., & Kovacevic, R. (2013). Parametric study on a coaxial multi-material powder fed laser deposition system. *Journal of Materials Processing Technology*, 213(9), 1594–1604.
96. Gockel, J., Sheridan, L., Koerper, B., & Whip, B. (2019). The influence of AM processing parameters on surface roughness and fatigue life. *Journal of Manufacturing Processes*, 44, 12–20.
97. Messaoudi, H., Naït-Said, R., & Trabelsi, M. (2018). Multi-objective optimization of selective laser melting process using genetic algorithms. *The International Journal of Advanced Manufacturing Technology*, 96, 2897–2911.
98. Peng, T., et al. (2018). Sustainability of AM: An overview on its environmental impact. *Journal of Cleaner Production*, 158, 116–129.
99. Bikas, H., Stavropoulos, P., & Chryssolouris, G. (2016). AM methods and modeling approaches: A critical review. *The International Journal of Advanced Manufacturing Technology*, 83, 389–405.
100. Sutton, R.S., & Barto, A.G. (2018). *Reinforcement Learning: An Introduction* (2nd ed.). Cambridge, MA: MIT Press.

101. Tang, M., & Pistorius, P.C. (2017). Oxides, porosity, and fatigue performance of AlSi10Mg parts produced by selective laser melting. *International Journal of Fatigue*, 94, 192–201.
102. Li, X., Roberts, M., & Fang, X. (2020). Reinforcement learning-based control of laser power in AM. *Robotics and Computer-Integrated Manufacturing*, 63, 101890.
103. Seifi, M., et al. (2016). Overview of materials qualification needs for metal AM. *JOM*, 68, 747–764.
104. Kamath, C. (2016). Data mining and statistical inference in selective laser melting. *The International Journal of Advanced Manufacturing Technology*, 86, 1659–1677.
105. Liu, Y., et al. (2015). Investigation into spatter behavior during selective laser melting of AISI 316L stainless steel powder. *Materials & Design*, 87, 797–806.

Chapter 3

Recent advancements in additive manufacturing

Anant Bhardwaj, Ankit Tyagi, Amit Mandal, and Rohit Sahu

3.1 INTRODUCTION

Additive manufacturing (AM, also known as 3D printing), because of its many advantages over traditional techniques, such as quick production, personalized products, lower labor and waste costs, accuracy, and precision, has been seen as having the potential to play a big part in the design and construction of the built environment [1]. In addition to enabling the creation of unusual structures that are not feasible with traditional construction methods, AM technologies can offer quick housing solutions in emergency situations [2,3]. To address the skilled labor shortage and prevent dangerous conditions at building sites, developed nations have been making large investments in automation of the construction industry. Even though AM methods have been around for over 30 years, there hasn't been much advancement in terms of automation for building applications [4].

This is due to the fact that large-scale construction applications cannot directly leverage AM techniques in their current state [5]. The development of sustainable new materials for construction applications, the achievement of ideal process parameters and conditions, the creation of computational modeling techniques, the addition of reinforcements and improvements to AM structures, and the development of new technologies remain urgent research issues despite recent advances in AM processes and technologies [6]. While some researchers have attempted to print structures using AM processes using traditional building materials [7], others have used new materials [8], such as recycled materials, geopolymer concrete, natural fiber-reinforced materials, and waste materials-reinforced construction materials.

Even though AM techniques have been enhanced to meet construction needs, materials that can integrate with the aforementioned processes and substitute traditional materials from the market still need to be developed. Because AM techniques need the material to be fluid enough to pass a delivery system, while rapid curing is necessary to assure structural integrity, researchers encounter challenges when trying to optimize process parameters. Extrudability, buildability, printer nozzle design, printing

DOI: 10.1201/9781003607052-3

speed and direction, aggregate sizes, ideal mixtures, and the inclusion of reinforcements in the material have all been the subject of research. However, the construction industry's standardization of AM procedures has not advanced to a point where it can be commercialized [6]. Through a methodical, in-depth, and comparative examination of the available literature, the writers of this article expand upon and enhance the earlier reviews, which are summed up in the following paragraphs. To highlight the constraints and barriers to the use of AM processes in the construction industry, as well as to provide guidance for future research areas for researchers, this study took into consideration both review papers and original research publications from reputable journals. Even though this field has a number of review articles, many of them are lacking in providing a multifaceted perspective and conversation. The papers of R. Buswell et al. [9], S. Ghaffar et al. [10], D. Delgado et al. [11], and J. Zhang et al. [12] are a few examples.

In their discussions of the material science involved in the process, G. D. Schutter et al. [13], V. Li et al. [14], and V. Mechtcherine et al. [15] did not go into great detail regarding the processes' development. On the other hand, F. Craveiro et al. [16] and A. Paolini et al. [17] provided the procedures in detail with little to no attention to the problems associated with the implementation of materials in 3DP processes. Finally, given the notable improvements in materials and methods that have been observed in the field in recent years, I. Perkins et al. [18], F. Bos et al. [19], and Y. Tay et al. [20] are regarded as out-of-date reviews.

The three most well-known AM techniques in the construction industry—contour crafting, concrete printing, and D-Shape—were reviewed by I. Perkins et al. [18]. There were reports of potential uses, cost analysis, and other advantages over traditional building techniques. The Eindhoven University of Technology's 3D concrete printing facility, along with other printing methods, was introduced by F. Bos et al. [19]. Potential research directions were discussed along with design, geometry, and process parameter issues. The development of 3D concrete printing technologies at the Singapore Centre for 3D Printing was the primary focus of Y. Tay et al.'s review of articles published between 1997 and 2016 [20].

A study of the fresh and hardened properties of mortar and concrete, as well as their impact on the mechanical and geometrical aspects of the structures, was given by R. Buswell et al. [9]. A review of cementitious and polymeric building materials, as well as the socioeconomic and environmental effects of AM techniques in construction, was published by S. Ghaffar et al. [10]. An article by D. Delgado et al. [11] reviewed AM in the construction sector, noted trends in AM materials and processes, and suggested future developments for AM applications. In their article, J. Zhang et al. [12] thoroughly examined the mechanical characteristics of 3D printable materials as well as material mix designs.

A study by Lu et al. [21] sought to methodically close the gap between the need for 3D printable cementitious materials and their research and development. In their analysis of AM applications in the construction industry, F. Craveiro et al. [16] described the steps involved in digital construction, the materials that are available, and their perspectives. AM methods and systems in the construction industry were recently examined by A. Paolini et al. [17]. A more comprehensive assessment of the literature on the use of metal AM technologies was given by C. Buchanan et al. [22], who also touched on several applications in the construction industry. A feasibility study on several drone-based technologies for masonry building of useful structures was conducted by S. Goessens et al. [23].

Recently, a comprehensive study was presented by S. Cai et al. [24] to assess the growth of automation in the high-rise building construction sector and to analyze the market in comparison to academic research. Three primary gaps between academia, the construction sector, and the robotics industry were noted, along with suggestions for closing them. A scientometric analysis and visualization of the literature on artificial intelligence in architecture, engineering, and construction was carried out by A. Darko et al. [25]. In order to validate the suggested framework, Rakha and Gorodetsky [26] conducted a case study after providing a thorough assessment of the use of unmanned aerial systems for built environment applications in automated building inspection.

In an essay devoted entirely to technical concerns pertaining to concrete 3D printing, G. D. Schutter et al. [13] added up the technological potential in terms of both the economy and the environment. Recent research on the properties and process features of 3D printable cement-based composites in both their fresh and hardened states was presented by V. Li et al. [14]. Although the mechanical properties of cementitious composite materials were covered in greater detail in this work, V. Mechtcherine et al. [15] provided a study that was nearly identical for extrusion-based techniques. A review article by D. Asprone et al. [27] focused specifically on reinforcement for concrete structures that were 3D printed. N. Roussel's review study [28] compiled existing research to determine the rheological specifications needed for concrete printing. This chapter discusses additive manufacturing in detail.

3.2 ADDITIVE MANUFACTURING

Additive manufacturing (AM) is a process of creating objects by adding material layer by layer based on a digital model. Unlike traditional subtractive manufacturing, which removes material from a solid block, AM builds components with minimal waste and allows for complex geometries that are difficult or impossible to achieve using conventional techniques. Talking about the classification of AM, it is broadly classified into various types, of

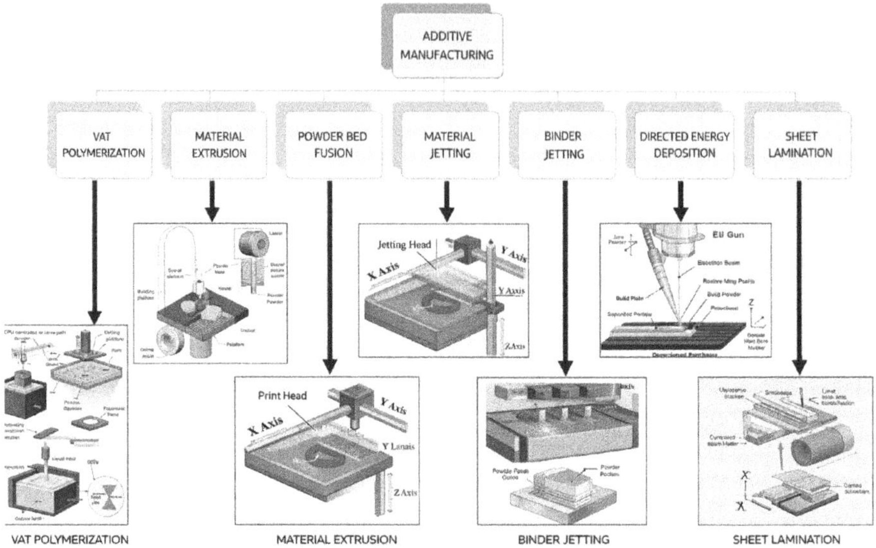

Figure 3.1 Additive manufacturing and its types.

which the first type is vat photopolymerization, the second type is material extrusion, the third type is powder bed fusion, the fourth type is material jetting, the fifth type is binder jetting, the sixth type is direct energy deposition, and the last type is the sheet lamination process. Figure 3.1 shows the brief classification of the modern AM process.

3.2.1 Vat photopolymerization

Vat polymerization is one of the prominent methods when it comes to the additive manufacturing process. This process uses a vat of liquid photopolymer resin, which is selectively cured (hardened) by ultraviolet (UV) light. The object is built layer by layer as light solidifies the resin in the required shape. Mainly, it uses several techniques for the process. The various subtypes of these vat polymerization processes are stereolithography (SLA), digital light processing (DLP), Vat photopolymerization, and continuous liquid interface production (CLIP).

Stereolithography (SLA) uses a laser to trace each layer in the resin, curing it as it goes. This results in high-precision parts with smooth surfaces. It is one of the earliest and most precise additive manufacturing (AM) technologies. It belongs to the Vat Photo polymerization category and is widely used for producing high-detail, smooth-finish prototypes and functional parts.

Ultraviolet light is one of the prominent methods for working with the solid layers. The UV lights harden the liquid hard the surface into photopolymer resin into several solid layers. The various steps involved in the

SLA process involves the initial preparation and setup, which includes the development of a 3D solid model with the help of CAD software and sliced them into various sublayers. These sliced sublayers are sent to the connected SLA printers containing vat- filled liquid resins. The next steps involve layer- by- layer curing. This is done when the UV light traces the shape of the layers. This exposure of light onto the liquid surface solidifies the liquid resin surface where needed. Once the surface gets solidify it moves downwards and the fresh resin takes its place and takes the space of the previous layers. The process continues until the entire object is completely filled. The final steps in the SLA includes post- processing, which involves removal of printed parts from the resin vat. Then the part is washed with isopropyl alcohol, where the excess resin is carried away and finally the material is strengthening by a UV curing step. The surface quality of the product may be obtained by processing an extra process involving sanding, painting, and coating.

Further, the printer used in the SLA process possesses various subtypes, which include top-down SLA printer. In this type of printers, the laser is used to set the resin vat and the part, and the build platform moves downward with the formation of layers. They are often used when a large volume is built. The second type of SLA printers are bottom-up SLA printers (inverted SLA printers). Here, in contrast with the first type, the interaction of the light source occurs below the resin vat also the platforms move at the upward direction instead of moving downward direction. It is generally used in the low- volume applications. It involves less use of resin hence it is cost- effective and used in a wide number of applications.

The advantages of SLA include its capability of producing parts with fine features deep up to 25 microns and providucing a smooth surface finish compared to FDM printing. It provides a variety of resin options involving castable, flexible, standard, tough, and high temperature resins, which makes it suitable for engineering, dental, jewelry, and medical industries. The method also possesses some limitations that include fragile parts, limited material strength, required post-processing (involving cleaning of surface, treatment with UV rays, sand treatment, and coating for durability), and lastly, the cost, which is higher than the FDM parts. The various key features of the SLA are extremely high accuracy with improved surface finish, the process application involves detailed prototype, dental models, and jewelry. The product generally involves custom aligners, various surgical instruments and prosthetic legs and hands. Talking about the jewelry, various products are made by wax prototypes casting. Apart from the engineering, automotive, and aerospace, SLA finds its application in eyewear, fashion accessory and miniatures parts which is difficult to produce by other methods. The process possesses certain disadvantage as the photopolymer produce is generally brittle. Most of the products require post-processing that involves cleaning and ultraviolent curing operation. Thus, SLA is one of the versatile 3D printing technologies for prototypes, miniatures, dental models, and intricate designs. but requires specialized resins, post-processing, and careful handling to maximize part durability.

The second form of the Vat polymerization is the DLP. The main process in this type is the use of digital projectors, which creates all the layers of the object rapidly. The DLP process involves the following steps: the first step is preparation and setup, which includes the slicing of 3D model into different layers and sending them to printers. The build platform is dipped into the vat, which is filled with photosensitive liquid resin. The second step is light projection and layer curing, which includes a digital projector that flashes the image of the layer onto the resin. Then this resin is exposed to the UV light which solidifies the exposed part of the resin. This solidified/build part moves upward which is displaced by the new parts and the cycle continues. The post-processing involves washing the object in isopropyl alcohol (IPA), which removes excess resin, the process part is then cured with UV light, and special treatment involving painting, sanding, or coating may also be given based on the requirement. Thus, the key difference between the SLA and the DLP process is that the former uses ultraviolet light and the later uses digital projector. Also, the DLP is faster and more efficient than SLA, as it cures entire layers in one go, but SLA offers higher detail and smoother surfaces due to its laser precision.

The advantages of the DLP method include rapid processing and a detailed and accurate fabrication that involves finishing up to 25 microns, which is ideal for jewelry and medical (dental) applications. DLP supports various materials that are standard, tough, flexible, castable, and dental resins, and it requires less liquid resin than SLA. Considering the disadvantages of the DLP process, the main limitations involve limited build volume, pixilation effect, high post-processing requirements, and resin sensitivity that becomes brittle due to higher exposure to sunlight and heat. Some common applications of the process involve jewelry, dental and medical, engineering and product design, miniature part, and automotive and aerospace applications.

Another type of vat polymerization process is the CLIP, which increases the speed of 3D printing. It includes 3D manufacturing in a similar way, but the difference is in the manufacturing method. It uses a continuous process that tends to reduce the number of layers and increases the speed of production. In this process, the object is pulled from the liquid resin, which provides the mentioned advantage to the process.

In this process, there is a continuous layer formation between each part, thus making it faster and innovative compared to SLA and DLP methods. This method is mostly used in aerospace, medical, automotive, and consumer products. The steps involved in the clip process involve the building platform of resin vat where the liquid photopolymer filled the vat and then the platform is kept above the resin. In the second step this vat is an oxygen window is used from where the UV light is passed through the DLP process which creates a dead zone which is preliminary a layer of uncured resin lying between the window and the growing part. In the next portion, the printed object is gradually lifted upward and is continuously cures the new layers with uninterrupted part growth. The oxygen used in the process ceases the initial part of the resin from solidifying, allowing the smooth material flow with high-speed printing. The final step involves post-processing in which isopropyl is used to wash and remove the excess resin. Then, the strengthening and revival of the mechanical property of the object is done by heat- based curing process. Lastly the additional finishing based on paint, sand and coating is done for additional finish. The process is common and uses extremely fast printing, involves seamless parts where no lines between the layers exists, the produce part with CLIP methods are stronger and durable with high mechanical strength. The most important application of CLIP method is in automotive, aerospace, medical and sports equipment manufacturing which makes it suitable for mass production. The process involves some limitations which involves the inclusion of expensive equipment and resin, the limitation of printer makes it more difficult to reach. The requirement of expertise is there and the post- processing treatment of the process is mandatory. The other advantage of the process consists of speedy resin-based fabrication process with high surface quality and strength also it eliminates the layer lines, resulting in efficient production. The industrial application of the CLIP process involves the manufacturing of durable parts, with lightweight parts along with strong and heat resistant aerospace components.

3.2.2 Material extrusion (FDM/FFF)

The next prominent method of 3D printing and additive manufacturing is material extrusion. It is one of the most common and accessible 3D printing methods. The methods predominantly include a thermoplastic filament, which is heated and extruded through a nozzle to form layers of different materials. The process consists of three main steps, which involves the feeding of filament spool into the printer head. The spool mainly consists of PLA, ABS, and PETG. The printer head moves along the specified path, which results in the layer by layer melting and deposition of the material. These heated layers of material cool and solidify to form a solid structure. The advantage of the process is the availability and structural rigidity of the PLA part. The limitations of the process are it possesses lower resolution when compared with other AM process also the manufactured part contains considerable visible layers, and wrapping issues are also most significant. FDM and FFF are the most common methods for material extrusion.

3.2.3 Powder bed fusion (PBF)

Powder bed fusion is one of the prominent methods used for 3D printing and the additive manufacturing process. In this process, a thin layer of substance, in the form of polymer or metal, is thoroughly spread over the surface, and it fuses with a laser beam and an electronic beam to form an additive manufacturing part. There are various subtypes of PBF process of which the main three are SLS, SLM, and EBM processes. The name SLS stands for selective laser sintering, in this process the polymer powder is sintered using the laser and the structure is created. In this process, there is no need for the support structure as the powder itself acts as a support. The second method is SLM which is abbreviated as selective laser melting; it is also called as direct metal laser sintering process. This process is often used for the additive manufacturing (3D printing) of metal. The process involves the sintering of material with a high-power laser that results in the melting and mixing of materials. The third type of process involves electron beam melting which is similar to SLM. The only thing which differentiates it from the SLM process is that it involves the use of electron beam instead of laser beam. The process finds its application in aerospace – grade metal like titanium. The advantage of the process involves production of high strength parts with complex geometries, and the process is ideal for aerospace, medical implants and aerospace parts. Talking about the limitations, it involves the use of expensive equipment with high energy consumption equipment. Also, the post-processing process is expensive and takes time.

3.2.4 Material jetting (MJ)

The next type of additive manufacturing process is material jetting. The process of material jetting possesses a similar nature compared to the ink-jet printing but uses small droplets of photopolymer or wax-like material, instead of ink, which is deposited layer- by- layer and are cured using UV light. The subtype of the process involves polyjet printing, which is used to produce fine and detailed parts. Another type of material jetting is nanoparticle jetting which uses a nanoparticles suspended in liquid. These nanoparticles are sintered to form a solid metal part made of metals. The advantage of the process involves a smooth surface finish with multi material and full-color model. The limitations of the process are its expense and the brittle nature of the parts, which makes it unsuitable for the structure applications.

3.2.5 Binder jetting (BJ)

The next type of additive manufacturing is binder jetting. In this method, liquid binders are readily used, which selectively binds the powder particles together. In this process, first the binding agent is deposited onto a thin layer of powder through a print head. The process is continuously repeated till the object is completely repeated. The excess is powder is completely removed, and the part is then cured. The process also involves post- processing including sintering of parts and providing the strength to final product. The key advantage of the process involves cost- effectiveness and large production runs. The products are adjustable in metal ceramics, and sand- casting moulds. The limitation of the process involves post-processing, such as sintering or infiltration also the mechanical strength of the powder is bit low compared with PBF.

3.2.6 Directed energy deposition (DED)

DED stands for directed energy deposition. This method finds its processing similar to welding, but unlike the conventional welding techniques, the proposed method uses a laser, electron beam, or plasma arc, which are high- energy sources require to melt and deposit material as it is being printed. The various subtypes of the DED process involves LENS process, which is abbreviated as laser engineered net shaping. The process creates a complex metal part with a laser and a powder metal part. Another subtype of the process is WAAM, which is abbreviated as wire arc additive manufacturing which involves the melting of metallic wire by virtue of electric arc. WAAM is generally used for the production of large- scale parts of shipbuilding and aerospace industry. The process is ideal for improving and updating metal parts, creating large structural components but the limitations involve the production of less precise parts and need machining and finishing at tight tolerances.

3.2.7 Sheet lamination (SL)

Another form of the additive manufacturing process, which is widely used, is sheet lamination. This process simply binds the different sheets of metal together to form a solid object. There are two subtypes of this process, which include laminated object manufacturing (LOM) and ultrasonic additive manufacturing (UAM). The former uses the application of laser to cut the metal plastic sheet, which is bonded with the adhesive-coated paper, and the latter uses ultrasonic welding, which acts as a bond between the thin metal sheet. The advantage includes a fast and cheap process made for large models. The process is applicable in rapid prototyping and visual model but the process reduces the mechanical strength and requires large post-processing of the additive manufacturing part (Table 3.1).

Table 3.1 Additive manufacturing process

Sr. no	*Types of process*	*Type of material*	*Advantage*	*Disadvantage*
1	Vat polymerization (SLA, DLP, CLIP)	Liquid resin	• Precise part • High detailing	• Brittle materials • Requirement of post-processing
2	Material extrusion (FDM/FFF)	Thermoplastics	• Prototyping • Functional parts	• Lower resolution • Layer lines available
3	Powder bed fusion (SLS, SLM, DMLS, EBM)	Polymers, Metals	• Strong mechanical • Complex parts	• Expensive • Slow • Requirement of post-processing
4	Material jetting (MJ, poly jet)	Photopolymers	• Full-color • High-resolution models	• Brittle parts, high cost

(Continued)

Table 3.1 (*Continued*) Additive manufacturing process

Sr. no	*Types of process*	*Type of material*	*Advantage*	*Disadvantage*
5	Binder jetting (BJ)	Metals, ceramics, sand	• Mass production • Casting molds	• Requires post-processing
6	Direct energy deposition (DED)	Metals	• Repairing or updating existing part	• Less precise, requires machining
7	Sheet lamination (LOM,UAM)	Paper, metal	• Low-cost large models • Fast processing	• Weak mechanical properties

3.3 KEY ASPECTS OF ADDITIVE MANUFACTURING

As observed from this chapter, various forms of additive manufacturing are done through various 3D printed process. This 3D printing is done with the different type of processes and thus requires different material for the processing. The different type of materials includes PLA, which stands for polylactic acid, ABS, which is abbreviated as acrylonitrile butadiene styrene, polyethylene terephthalate glycol, nylon and the resin falls in the category of polymers. Talking about the metals, the different metals are titanium, aluminium, stainless steel, cobalt and chromium. Apart from metal and polymers, ceramics like alumina and zirconium plays a vital role in the 3D printing applications. Moreover, latest applications have highlighted the use of composites, which mostly contain carbon fiber-reinforced polymers as the prominent printing materials. The overall usefulness of this material is for the fabrication of the lightweight structure with complex geometry, the other focus is in the green manufacturing where reduced material waste and low energy consumption occurs. The process fosters prototyping with rapid market needs thus, reducing the major requirement of heavy tools and assembly units. The applications involve but are not confined to the aerospace and automotive sectors due to the presence of lightweight components, medical and healthcare sector involving prosthetics and dental applications. The additive manufacturing is emerging rapidly in the area of construction where 3D printing buildings and infrastructure are taken into wide used in the remote areas where heavy construction instruments are difficult to reach. Apart from the construction area additive manufacturing is emerging in the area of consumer goods. They are mostly used in footwear, eyewear and fashion items. Finally, the additive manufacturing and has replaced the major mechanical application generally in the area of tooling, jigs, and fixtures. Table 3.2 shows different 3D printers, their features, and their process of application, and Figure 3.2 highlights the major areas of application of the additive manufacturing process.

Table 3.2 Types of 3D printers

Sr. no	*Company*	*Model specification*	*Key features*	*Process*
1	Carbon	M3 max	• Large volume • Industrial-grade components	CLIP
2	Carbon	L1 printer	• High-speed production • Mass production	CLIP
3	Carbon	M2	• Compact • Less working parts • Suitable for engineering and medical applications	CLIP
4	Formlabs	Form 3+	• Professional-grade • Reliable and equipped • High detailing	SLA
5	Anycubic	Photon mono X	• Affordable, • Fast printing, • contains monochrome LCD	SLA
6	Elegoo	Mars 4 ultra	• Economic • High resolution	SLA
7	Peopoly	Phenom XXL	• Heavy equipped • Large build volume • Industrial applications	SLA

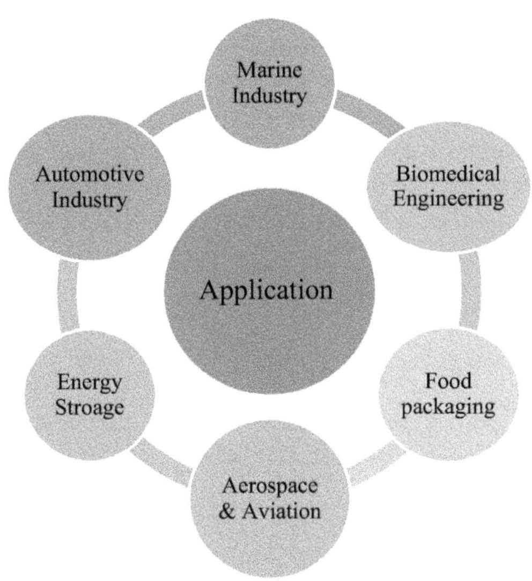

Figure 3.2 Application of additive manufacturing.

3.4 CONCLUSIONS

Additive manufacturing's remarkable qualities and adaptability have made it essential to many sectors. In the aerospace industry, its high strength-to-weight ratio improves fuel efficiency and aircraft performance. By using these materials to lighten vehicles, the automobile sector can increase fuel efficiency and lower pollution. Additive manufacturing process, which provides durability and resilience to environmental variables, is used in civil engineering to strengthen structures. It is used in equipment that improves athlete performance by making it stronger and lighter, which benefits the sports business. Additionally, because of its biocompatibility and adaptability, additive manufacturing products are used in biomedical applications for implants and prosthetics. Additive manufacturing's ongoing development and application in a variety of industries highlights itsimportance in contemporary innovation and technology.

A family of materials known for its exceptional qualities and broad range of applications are carbon-integrated polymer composites. Numerous production techniques are used, each with unique advantages. These technologies include 3D printing, electrospinning, melt and solution mixing, and in situ polymerization. Thermomechanical, optoelectronic, and magnetic features, as well as a high density, enhanced strength, huge surface area, and greater elastic modulus, make these materials attractive for innovative uses. The evolution of polymer composite synthesis and production is explored in depth in this article. Biomedical engineering, aerospace, the automobile industry, and energy storage are just a few of the several additional sectors that use these cutting-edge materials. The field's prospects and problems are brought to light via discussions of possible advances and planned research initiatives. Future developments in hybrid systems and the functionalization of carbon materials are explored. However, issues with interfacial compatibility, scalability, cost-effective manufacturing, and environmental sustainability must be resolved before widespread use is feasible.

REFERENCES

1. K. Rane, M. Strano, A comprehensive review of extrusion-based additive manufacturing processes for rapid production of metallic and ceramic parts, *Adv. Manuf.* 7 (2019) 155–173, https://doi.org/10.1007/s40436-019-00253-6.
2. B. Khoshnevis, Toward total automation of on-site construction - an integrated approach based on contour crafting, *Proceedings of the 20th International Symposium on Automation and Robotics in Construction (ISARC 2003)*, September 21–24, 2003, Eindhoven, The Netherlands. Eindhoven: International Association for Automation and Robotics in Construction, pp. 61–66, https://pdfs.semanticscholar.org/eb6c/bdbf29a95f6f19d4fd0bcb8294 57f163dfb9.pdf.

3. B. Khoshnevis, Automated construction by contour crafting - related robotics and information technologies, *Autom. Constr.* 13 (2004) 5–19, https://doi.org/10.1016/j.autcon.2003.08.012.
4. F. Hamidi, F. Aslani, Additive manufacturing of cementitious composites: materials, methods, potentials, and challenges, *Constr. Build. Mater.* 218 (2019) 582–609, https://doi.org/10.1016/j.conbuildmat.2019.05.140.
5. B. Khoshnevis, M.P. Bodiford, K.H. Burks, E. Ethridge, D. Tucker, W. Kim, H. Toutanji, M.R. Fiske, Lunar contour crafting - a novel technique for ISRU-based habitat development, *43rd AIAA Aerospace Sciences Meeting and Exhibit-Meeting Papers*, January 10–13, 2005, Reno, Nevada, USA. Reston, VA: American Institute of Aeronautics and Astronautics, pp. 7397–7409, https://doi.org/10.2514/6.2005-538.
6. A. Bhardwaj, S.Z. Jones, N. Kalantar, Z. Pei, J. Vickers, T. Wangler, P. Zavattieri, N. Zou, Additive manufacturing processes for infrastructure construction: a review, *J. Manuf. Sci. Eng.* 141 (2019) 1–13, https://doi.org/10.1115/1.4044106.
7. S.C. Paul, Y.W.D. Tay, B. Panda, M.J. Tan, Fresh and hardened properties of 3D printable cementitious materials for building and construction, *Arch. Civ. Mech. Eng.* 18 (2018) 311–319, https://doi.org/10.1016/J.ACME.2017.02.008.
8. M. Mohammad, E. Masad, T. Seers, S.G. Al-Ghamdi, Properties and microstructure distribution of high-performance thermal insulation concrete, *Materials* 13 (9) (2020), https://doi.org/10.3390/ma13092091.
9. R.A. Buswell, W.R. Leal de Silva, S.Z. Jones, J. Dirrenberger, 3D printing using concrete extrusion: a roadmap for research, *Cem. Concr. Res.* 112 (2018) 37–49, https://doi.org/10.1016/j.cemconres.2018.05.006.
10. S.H. Ghaffar, J. Corker, M. Fan, Additive manufacturing technology and its implementation in construction as an eco-innovative solution, *Autom. Constr.* 93 (2018) 1–11, https://doi.org/10.1016/j.autcon.2018.05.005.
11. D. Delgado Camacho, P. Clayton, W.J. O'Brien, C. Seepersad, M. Juenger, R. Ferron, S. Salamone, Applications of additive manufacturing in the construction industry – a forward-looking review, *Autom. Constr.* 89 (2018) 110–119, https://doi.org/10.1016/j.autcon.2017.12.031.
12. J. Zhang, J. Wang, S. Dong, X. Yu, B. Han, A review of the current progress and application of 3D printed concrete, *Compos. A: Appl. Sci. Manuf.* 125 (2019) 105533, https://doi.org/10.1016/j.compositesa.2019.105533.
13. G. De Schutter, K. Lesage, V. Mechtcherine, V.N. Nerella, G. Habert, I. Agusti-Juan, Vision of 3D printing with concrete — technical, economic and environmental potentials, *Cem. Concr. Res.* 112 (2018) 25–36, https://doi.org/10.1016/j.cemconres.2018.06.001.
14. V. Li, F. Bos, K. Yu, W. McGee, T.Y. Ng, S.C. Figueiredo, K. Nefs, V. Mechtcherine, V.N. Nerella, J. Pan, G. van Zijl, J. Kruger, On the emergence of 3D printable engineered, strain hardening cementitious composites (ECC/SHCC), *Cem. Concr.Res.* 132 (2020) 106038, https://doi.org/10.1016/j.cemconres.2020.106038.
15. V. Mechtcherine, F.P. Bos, A. Perrot, W.R.L. da Silva, V.N. Nerella, S. Fataei, R.J.M. Wolfs, M. Sonebi, N. Roussel, Extrusion-based additive manufacturing with cement-based materials – production steps, processes, and their underlying physics: a review, *Cem. Concr. Res.* 132 (2020) 106037, https://doi.org/10.1016/j.cemconres.2020.106037.

16. F. Craveiro, J.P. Duarte, H. Bartolo, P.J. Bartolo, Additive manufacturing as an enabling technology for digital construction: a perspective on construction 4.0, *Autom. Constr.* 103 (2019) 251–267, https://doi.org/10.1016/j.autcon.2019.03.011.
17. A. Paolini, S. Kollmannsberger, E. Rank, Additive manufacturing in construction: a review on processes, applications, and digital planning methods, *Addit. Manuf.* 30 (2019) 100894, https://doi.org/10.1016/j.addma.2019.100894.
18. I. Perkins, M. Skitmore, Three-dimensional printing in the construction industry: a review, *Int. J. Constr. Manag.* 15 (2015) 1–9, https://doi.org/10.1080/15623599.2015.1012136.
19. F. Bos, R. Wolfs, Z. Ahmed, T. Salet, Additive manufacturing of concrete in construction: potentials and challenges of 3D concrete printing, *Virtual Phys. Prototyp.* 11 (2016) 209–225, https://doi.org/10.1080/17452759.2016.1209867.
20. Y. Tay, B. Panda, S. Paul, N. Noor, M. Tan, K. Leong, 3D printing trends in building and construction industry: a review, *Virtual Phys. Prototyp.* 12 (2017) 261–276, https://doi.org/10.1080/17452759.2017.1326724.
21. B. Lu, Y. Weng, M. Li, Y. Qian, K.F. Leong, M.J. Tan, S. Qian, A systematical review of 3D printable cementitious materials, *Constr. Build. Mater.* 207 (2019) 477–490, https://doi.org/10.1016/j.conbuildmat.2019.02.144.
22. C. Buchanan, L. Gardner, Metal 3D printing in construction: a review of methods, research, applications, opportunities and challenges, *Eng. Structure.* 180 (2019) 332–348, https://doi.org/10.1016/j.engstruct.2018.11.045.
23. S. Goessens, C. Mueller, P. Latteur, Feasibility study for drone-based masonry construction of real-scale structures, *Autom. Constr.* 94 (2018) 458–480, https://doi.org/10.1016/j.autcon.2018.06.015.
24. S. Cai, Z. Ma, M.J. Skibniewski, S. Bao, Construction automation and robotics for high-rise buildings over the past decades: a comprehensive review, *Adv. Eng. Inform.* 42 (2019) 100989, https://doi.org/10.1016/j.aei.2019.100989.
25. A. Darko, A.P.C. Chan, M.A. Adabre, D.J. Edwards, M.R. Hosseini, E.E. Ameyaw, Artificial intelligence in the AEC industry: scientometric analysis and visualization of research activities, *Autom. Constr.* 112 (2020) 103081, https://doi.org/10.1016/j.autcon.2020.103081.
26. T. Rakha, A. Gorodetsky, Review of unmanned aerial system (UAS) applications in the built environment: towards automated building inspection procedures using drones, *Autom. Constr.* 93 (2018) 252–264, https://doi.org/10.1016/j.autcon.2018.05.002.
27. D. Asprone, C. Menna, F.P. Bos, T.A.M. Salet, J. Mata-Falcón, W. Kaufmann, Rethinking reinforcement for digital fabrication with concrete, *Cem. Concr. Res.* 112 (2018) 111–121, https://doi.org/10.1016/j.cemconres.2018.05.020.
28. N. Roussel, Rheological requirements for printable concretes, *Cem. Concr. Res.* 112 (2018) 76–85, https://doi.org/10.1016/j.cemconres.2018.04.005.

Chapter 4

Industry 6.0 and additive manufacturing

Applications and future directions

Parveen Sihmar, Manjeet Chahar, Deepak Dahiya, Upender Punia, and Akash Ahalwat

4.1 INTRODUCTION

Every advancement in the technology of manufacturing processes has come with benefits regarding process optimization, networking, and speed [1]. The leap to Industry 6.0 signifies that we have transitioned to the next epoch for technological progress. It is indicative of a shift in the mash-up of technological development and human experiences to harness new opportunities in the reconfiguration of manufacturing [2]. This progress merges human intuition with core digitization and cyber-physical systems, referred to as Industrial 6.0. It is distinct from other approaches due to its principles. Additive manufacturing is at the top of this revolution because it uses digital models to create intricate parts by addressing them one layer at a time [3]. This results in a reduction in the amount of wasted material, as opposed to traditional subtractive manufacturing methods [4]. This technology allows companies to integrate specific parts into their low- and high-volume flexible production systems. Additive manufacturing helps reduce operational expenditures while maintaining efficiency by solving the production problems of high cost and low machine capability [5]. Moreover, it contributes to the achievement of sustainability goals by reducing harmful waste and adhering to the core principles of the circular economy. Combining big data, machine learning, artificial intelligence, and the Internet of Things in the design for Industry 6.0 increases the flexibility in the manufacturing processes [6]. To achieve this goal, it combines the human knowledge of solving complex problems with automated systems. Industry 6.0 emphasizes the use of additive manufacturing processes for the creation of self-optimizing structures that aid in achieving sustainable goals. It is similar to the traditional approach of focusing on the latest technologies and inventions, but is different in essence [7]. The efficacy of machine learning algorithms is manifested in the increase in productivity from the additive manufacturing devices through the optimization of the material flow, the printer control settings, and the processing of

DOI: 10.1201/9781003607052-4

large data sets [8]. This occurs by the improvement of operational efficiency. In the particular case of creative factory production, it surfaces in the case of human interaction with machine data and sensors, where system analysis of operational data is constantly optimized to produce outputs that meet demand and increase productivity [9]. This is the case, especially for the creative factory production industry. The limited tooling capacities of traditional production processes made bulk personalization economically and temporally expensive [10]. At the same time, the additively manufactured technology allows for quick alterations in designs minimizing the production of personalized items on a large-scale. These include the fashion sector, which makes clothes and makeup; the automotive sector, which produces custom-tailored PCs; and medical implants, which health facilities provide [11]. Combining AMtechnologies alongside Industry 6.0 practices creates a new milestone in the production stage that commits to sustainable, effective, and people-centered solutions [12]. Together with the technologies for additive manufacturing, the applications of Industry 6.0 make it possible to construct flexible manufacturing systems that address consumers' requirements without compromising on the environmental impacts of production processes [13].

4.2 EVOLUTION

Manufacturing and production capabilities have been expanded, and structural economic changes have occurred due to successive technological evolutions in the history of industrial development [14]. The industry's operations have been essentially altered by each shift from mechanical work in Industry 1.0 to digitized and networked production facilities in Industry 4.0 [15]. The upcoming Industry 6.0 will facilitate the industrial transformation, establishing flexible, adaptive production systems prioritizing human factors [16]. This section discusses the evolution of Industry 6.0 as shown in Figure 4.1, from its inception through sequential industrial revolutions, and an analysis of the current intelligent production capability to ensure future readiness.

4.2.1 Industry 1.0 – the birth of mechanization

Discoveries of steam-powered technologies combined with mechanical machinery marked the beginning of Industry 1.0 in the late 18th century. During this revolutionary era, human history underwent a significant transformation as production transitioned from farms and handcraft creation to massive factory operations [17]. With James Watt's pioneering efforts, enterprises received self-generating power through steam engines, allowing them to expand textile production while extracting larger industrial amounts of coal and iron [18].

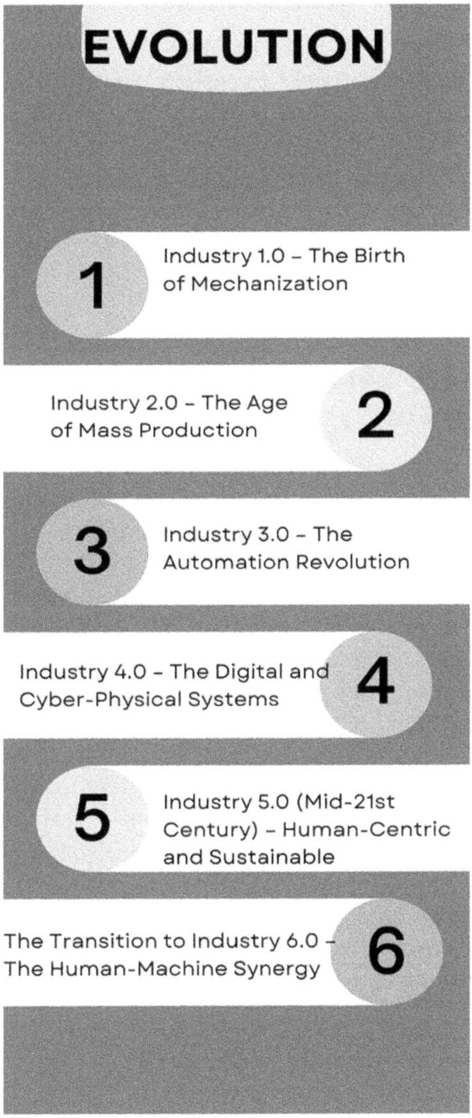

Figure 4.1 Evolution of Industry 6.0.

4.2.2 Industry 2.0 – the age of mass production

Industrial revolution innovations from the early twentieth century evolved to Industry 2.0 with the critical adoption of electric-powered assembly lines [19]. Production methods fielded through Henry Ford's assembly line achieved commercial expense reduction while promoting operational efficiency through selective component installation.

4.2.3 Industry 3.0 – the automation revolution

Computer technologies and automation served as the main drivers of industrial expansion in the latter twentieth century. Industrial operations achieved improved productivity through the implementation of robotic systems along with PLC systems and CAD applications [20]. Machines with repetitive abilities reduced human workforces when performing factory tasks throughout this period. Managers now possess superior control systems because of information technology. The consolidation of industrial systems to raise productivity levels became possible after implementing network-based industrial systems.

4.2.4 Industry 4.0 – the digital and cyber-physical systems

The early part of the 21st century introduced Industry 4.0 through physical-digital system combinations that created unprecedented production technologies. Standards from IoT to cloud computing to big data analytics to cyber-physical systems form the base for Industry 4.0 [21]. The current stage utilizes networked machinery, sensors, and autonomous devices alongside real-time information transmission methodologies. Highly efficient, end-to-end self-optimizing systems become possible through the joint power of AI, machine learning, and robotics capabilities [22]. Industry figures benefit from on-demand flexibility, which leads to fewer operational delays and enhanced production outcomes. Arising from Industry 4.0 advancements lies the next-level development within additive manufacturing (AM) and its capabilities in producing complex parts directly from digital files while decreasing the waste rate [23]. A window into Industry 4.0's potential is restricted by its dependence on factory-based manufacturing and isolated data management systems. Automation has boosted system efficiency, but most operations remain structured as a reactive framework that harmonizes human workers poorly with modern technological systems.

4.2.5 Industry 5.0 (mid-21st century) – human-centric and sustainable

Industry 5.0 illustrates a manufacturing methodology that emphasizes sustainable human-machine collaboration. Industry automation's fundamental concept was Industrial 4.0; however, Industry 5.0 places greater emphasis on environmentally conscious technological advancements, responsible AI practices, and custom solutions [24]. Robotic tools facilitate collaboration between human operators and robots by enabling professionals to engage in innovative and creative work while robots execute repetitive tasks. The primary foundation for sustainability is the circular economy, renewable energy, and waste reduction methods. Ethical AI methodologies enable transparency and equity toward employees and industry initiatives promoting social engagement and well-being [24]. This paradigm shift enables businesses and society to collaborate to pursue a sustainable and inclusive future by integrating ecological value, creativity, and technological advancement.

4.2.6 The transition to Industry 6.0 – the human-machine synergy

Industrial Revolution 6.0 established coupled human-machine activities in a continuous working system, surpassing conventional automation [25]. Combining artificial intelligence (AI) with cutting-edge robotics, big data, and human ingenuity produces a manufacturing system with flexible and tailored solutions and adaptability. Today, industrial manufacturing goes beyond production numbers to include innovative decision systems creating customized goods using environmentally friendly techniques. Under Industry 6.0, people work with machines in cooperative operations rather than automating whole human functions [26]. Robotic and robot systems can operate inside shared human-operated workspaces because of AI capabilities for analyzing vast amounts of data. The combination of human-led creativity, intuition, and problem-solving abilities functions alongside machine-operated precise data processing and rapid task execution in a collaborative work environment. The hallmark of Industry 6.0 exists in personalized manufacturing processes. The industrial revolution of Industry 6.0 combines data analytics with additive manufacturing technology to create batch production of custom-tailored products that match consumer requirements. The human-scale production of customized items becomes possible because technologies like AM allow businesses to create personalized medical devices and custom-designed consumer goods quickly and affordably [27]. The sustainability perspective stands out as a fundamental feature of Industry 6.0 operations. Industry 6.0 emerging technologies promote sustainable manufacturing through waste reduction while achieving maximum energy efficiency and advancing circular economy principles. Sector power in additive manufacturing bolsters resource utilization efficiency while decreasing material waste.

4.3 INTEGRATION OF INDUSTRY 6.0 WITH ADDITIVE MANUFACTURING

Traditional manufacturing methods transformed significantly by combining Industry 6.0 and additive manufacturing, which alters production thinking and execution operations [28]. Traditional production priority centered on efficiency, mass production, and cost minimization, but Industry 6.0 established flexibility and customization alongside sustainability and human-centric design as primary objectives. Additive manufacturing fits perfectly into the Industry 6.0 framework through its capability to create unique customized products and complex geometric forms [29]. Additive Manufacturing teams up with Industry 6.0 technologies, including AI, IoT, as well as Big Data and advanced robotics, to transform modern manufacturing systems.

4.3.1 The role of additive manufacturing in Industry 6.0

3D printing technology structures products by sequentially depositing material layers from digital designs into end products. Traditional subtractive production techniques use material destruction by elimination, while additive manufacturing allows producers to create demanding objects that standard industrial methods cannot quickly produce [30]. Within the Industry 6.0 framework, AM technology is the primary enabler of personalized, scalable production systems and instant assembly solutions. AM is a primary strength in Industry 6.0 manufacturing by enabling mass customization. The industry has endured a dual challenge because mass production techniques conflict with product customization specifications. Industry 6.0 addresses challenges through innovative combinations of AI, IoT, and big data analytics, which allows manufacturers to adapt production flows during the process and personalize their goods for each customer [31]. Thanks to its customization capabilities, AM technology now delivers personalized implant solutions and one-of-a-kind automotive components in at-scale manufacturing volumes. Sustainability emerges as a primary advantage thanks to AM technology. Manufacturers utilize precise material requirements through AM processes to produce components with reduced waste output compared to standard manufacturing methods [32]. Advanced Manufacturing allows for the use of recycled resources, which promotes the growth of circular economies through product reuse, remanufacturing processes, and recycling activities. Industry 6.0 leverages these capabilities through sustainability initiatives to increase material efficiency while lowering production energy requirements.

4.3.2 The synergy between Industry 6.0 technologies and AM

Integration with Industry 6.0 systems is the sole method by which the tangible potential of additive manufacturing for developing complex, customized parts can be realized [33]. These technologies enable additive manufacturing to function as an intelligent manufacturing framework component.

4.3.2.1 Artificial intelligence (AI)

Artificial intelligence systems provide indispensable assistance to AM's capabilities. After numerous testing sessions, traditional manufacturing companies must optimize design through automated methods and manual procedures [34]. AI algorithms facilitate the optimization of design parameters by comprehensively analyzing various design variables to determine the optimal dimensions for additive manufacturing. Predictive maintenance of AM machines is more effective when artificial intelligence continuously monitors system health to identify machine failures before they occur [35]. Quality control automation is achieved by analyzing AI data from sensors, which verify the accuracy of printed object layers according to the specifications.

4.3.2.2 Internet of Things (IoT)

Continuous monitoring and real-time printing process control are made possible by the integration of IoT devices with AM systems. The data collected by IoT sensors during printing contains extensive information about temperature, humidity, and material properties so that the machine's configuration can be altered for better outcomes. Machines monitored in real-time by manufacturers help reduce downtime and enhance production efficiency while avoiding defects. IoT technology makes smart factories possible because integrated devices communicate automatically to improve production efficiency [36].

4.3.2.3 Big Data

The extensive data production from additive manufacturing enables manufacturers to use Big Data analytical methods to discover valuable insights toward process optimization [37]. Instrument measurements, automated equipment, and product tracking generate data that becomes the basis for optimizing design steps and future projection analysis while allowing producers to identify manufacturing performance issues. Big Data technology helps manufacturers make better demand forecasts, improving just-in-time production and minimizing required quantities of stock inventory [38]. Using forecasting based on Big Data delivers excellent value to the aerospace and automotive sectors because they have high inventory expenses and changing part requirements.

4.3.2.4 Advanced robotics and automation

Under Industry 6.0 manufacturing environments, robots benefit production by teaming up with human counterparts to complete tasks [39]. AM technology systems integrate robotic systems that execute material handling, part assembly operations, and post-processing procedures, including cleaning, curing, and finishing. Using robots means being independent and working alongside humans (robots). This promotes AM efficiency through robotic automation of monotonous workflows, releasing the human workforce for higher-level responsibilities such as product design and problem resolution [40]. Reliable real-time process adjustments are possible through robot assistance based on sensor signals, providing enhanced precision and reduced printing errors.

4.3.3 The benefits of integration: customization, flexibility, and sustainability

Manufacturers can achieve significant production benefits by combining sustainable practices, operational efficiency, and product customization with Industry 6.0 assets and additive manufacturing technologies [41].

4.3.3.1 Customization and personalization

According to the combination of AM technology and Industry 6.0, manufacturers may now create unique, customized items. Traditional manufacturing finds it challenging to strike a balance between standardized mass production and the development of personalized products [42]. On the other hand, thanks to additive manufacturing, manufacturers can construct customized car parts and medical equipment.

4.3.3.2 Increased flexibility

Through Industry 6.0, manufacturers can safely obtain the necessary tools to help them quickly adapt to changes in market demand and production chain disruptions [43]. Through real-time design modifications, predictive analytics, and rapid part production capabilities, manufacturers can stay competitive in changing market conditions. Aerospace companies prioritize quick prototyping and alterations to enhance adaptability. Implementing bioprinting for parts in local production facilities boosts supply chain flexibility and efficiency by reducing manufacturing time and distribution costs.

4.4 APPLICATIONS OF ADDITIVE MANUFACTURING INTEGRATING WITH INDUSTRY 6.0

The integration of additive manufacturing (AM) with Industry 6.0 technology has improved product design, production, and delivery in various industrial sectors, including aerospace, healthcare, and automotive, by combining human-machine interaction, advanced analytics, AI, and real-time communications [44]. The application of additive manufacturing with Industry 6.0 is shown in Figure 4.2.

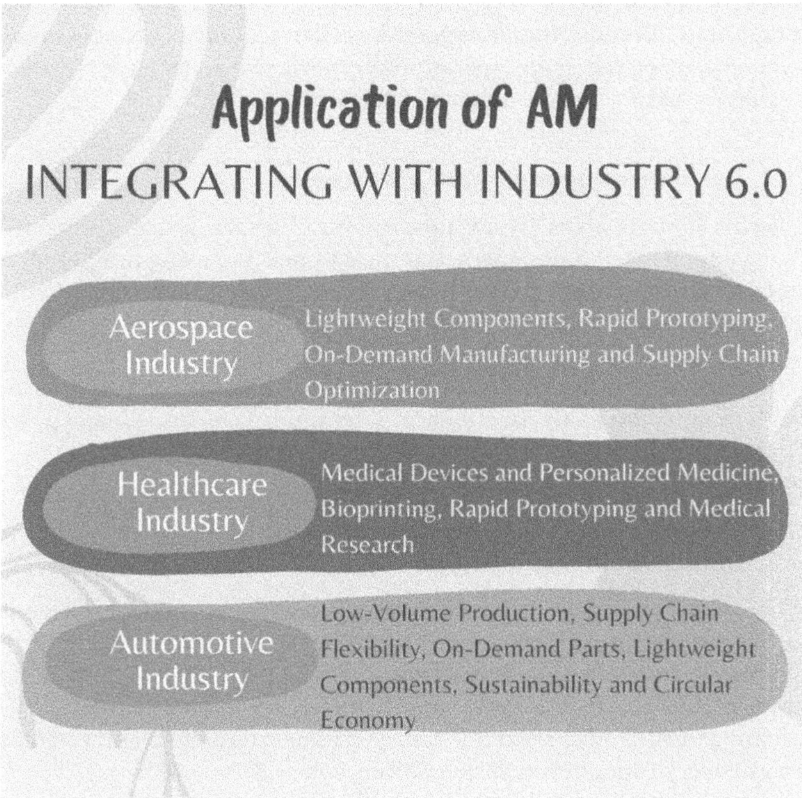

Figure 4.2 Applications of additive manufacturing integrating with Industry 6.0.

4.4.1 Aerospace industry

The aerospace sector benefits significantly from Industry 6.0 technology and additive manufacturing, utilizing casting, machining, and assembly techniques to manage intricate components from conventional manufacturing [45]. AM technology employs automated processes to fabricate intricate structures that reduce byproduct consumption and provide economical enhancements in performance.

4.4.1.1 Lightweight components

AM technology is mainly used in the aerospace sector to produce much lower-density parts. Since the beginning of aircraft construction, manufacturers have experimented with various methods to reduce their aircraft weight, increasing fuel efficiency [46]. Operators can create more intricate, lighter, and durable items using AM technology than those conventionally manufactured . Because they weigh far less than regular parts, these unique components meet the same operational criteria while using less fuel and having fewer adverse environmental effects.

4.4.1.2 Customization and rapid prototyping

The technology enables rapid prototyping for aerospace components, allowing engineers and designers to test new concepts using tangible prototypes quickly. This accelerates innovation by reducing the production time from initial design to commercial launch. Aerospace operators can modify specific parts using additive manufacturing to meet specific customer requirements and various aircraft designs [47]. This process optimization through design data analysis, enhances later improvements is made possible by implementing Industry 6.0 technologies, such as AI and IoT.

4.4.1.3 On-demand manufacturing and supply chain optimization

Aerospace businesses can now fulfill direct orders and eliminate the requirement to store extra components due to AM part manufacturing. The digital supply chain made possible by Industry 6.0 technologies, such as cloud computing and the IoT, allows manufacturers to maintain digital inventory management and carry out localized production as necessary, reducing overall delivery times and transportation costs [48].

4.4.2 Healthcare industry: personalized medicine and medical devices

4.4.2.1 Medical devices and personalized medicine

The integration of Industry 6.0 technologies and AM is causing a fundamental shift in the healthcare industry. Because AM technologies allow for patient-specific medical applications adapted to each patient's conditions, they offer significant value to the industry [49].

4.4.2.2 Personalized medical devices

Manufacturers use AM to create customized orthotics, prostheses, and medical implants that fit each patient's distinctive body structure. AM manufacturing technologies also produce customized replacement hip joints, dental implants, and hearing aids, giving patients more convenience and performance [50]. When combined with big data analytics systems, AI allows device manufacturers to develop customized products that deliver optimized medical functions for individual patients.

4.4.2.3 Bioprinting

Bioprinting is one of the most promising future uses of additive manufacturing in healthcare because 3D printers create tissues,organs, and complete anatomical structures from living cells. Robotic organ bioprinting shows promise as an upcoming solution that could transform how doctors produce customized organs for medical transplant programs [51]. Integrating AI and machine learning algorithms within the process enables the analysis of medical data, followed by assistance with functional biological structure generation.

4.4.2.4 Rapid prototyping and medical research

The use of AM advances development activities in medical devices and pharmaceutical research through rapid prototyping capabilities. Research teams benefit from 3D printing technology to manufacture physical representations of organs and body components for identification and experimental analysis. Through its quick prototyping features and testing capabilities, AM amplifies the pace at which researchers develop new treatments and medical therapies [52]. Precise models developed through AM-AI integration accelerate clinical trials by enabling medication and medical device pre-testing with human subject simulations before actual human clinical trials.

4.4.3 Automotive industry: customization, performance, and sustainability

The automotive field is a major regional adopter of additive manufacturing, achieving expanded potential through Industry 6.0 systems.

4.4.3.1 Low-volume production and customization

The automotive sector uses additive manufacturing techniques for building customized components in low-volume production runs directed toward luxury vehicles and high-performance cars, including electric vehicles [53]. The creation of custom parts traditionally demanded extensive mold and tooling resources, which limited the affordable production of small batches. The manufacturing process with AM technology bypasses expensive mold requirements to produce customized parts, thus shortening market entry timelines and saving costs. Industry 6.0 technology integration through AI and data analytics enables the customization of parts according to specific performance needs, thereby delivering personalized customer experiences.

4.4.3.2 Supply chain flexibility and on-demand parts

Through AM, manufacturers can fulfill automotive spare part requirements as needed with less dependence on stockpiled inventory. Managing significant inventories when they allay hundreds of components throughout multiple regions costs businesses time and money. Manufacturers can generate locally needed spare parts by implementing digital supply chain methods in Industry 6.0 and AM, reducing inventory expenses and delivery times [54]. The supply chain operates more effectively through this method because the correct components become available precisely when and where they should be while lowering costs.

4.4.3.3 Lightweight components and performance

AM adds lightweight design options to the automotive sector and aerospace applications. Engineers can employ AM to create parts that combine lower weight and increased strength, which improves performance and fuel efficiency. Manufacturers use AM to create lightweight brackets with bumpers and engine components to reduce car weight and increase energy efficiency regulations [55].

4.4.3.4 Sustainability and circular economy

Companies in the automotive industry use additive manufacturing to drive sustainability by reducing material waste. Conventional manufacturing practices lead to extensive material waste, which differs from additive manufacturing because it uses only the needed material. Through strategic integration of Industry 6.0 technologies, AI, and Big Data, the manufacturing process can be optimized for minimum material waste [56]. AM benefits the circular economy through its capability to design components that both simplify end-of-life disassembly and support material recycling operations, which build a sustainable production framework.

4.4.4 Other industries: consumer goods, construction, and more

Beyond the aerospace, healthcare, and automotive sectors, integrated applications from AM and Industry 6.0 technologies show potential solutions for many other industries, including consumer goods, construction, electronics, and fashion.

4.4.4.1 Consumer goods

People can receive customized products through AM technology, including tailor-made footwear and eyewear, personalized jewelry, and skin-tailored apparel. By uniting Industry 6.0 technology with real-time consumer data, manufacturers can modify products according to individual style and preference demands [57].

4.4.4.2 Construction

The construction industry is becoming more interested in 3D printing technologies that help create structural elements ranging from basic components to entire architecturally unique buildings. Large-scale AM manufacturing technologies enable operators to build horizontal surfaces, support foundations, and design unique architectural pieces simultaneously, minimizing construction time and financial expenses [58].

4.4.4.3 Electronics

The additive manufacturing process can generate unique electronic devices and develop customized prototypes. Through 3D printing, engineers generate electronic circuit prototypes, which lead to rapid testing and optimization procedures before starting production.

4.4.4.4 Fashion

Fashion industry leaders create 3D printed alternatives, including clothing items, sneakers, and wearable accessories. AM creates intricate product designs through its manufacturing technology that conventional production methods cannot replicate [59]. Potential Industry 6.0 tools allow designers to quickly create tailored versions of fashion products that meet fluctuating market demands.

4.5 FUTURE SCOPE AND CHALLENGES

Industry 6.0 partnerships with AM technologies introduce a fundamental transformation in modern manufacturing systems. As we move towards the future, we will face many growth and innovation possibilities, but we must also solve numerous issues to maximize these technologies' full potential [60]. This segment explores the future boundaries of Industry 6.0 and AM by highlighting possible advances and applications. It also explains the fundamental barriers that must be overcome to make sustainable manufacturing growth possible.

4.5.1 Future scope of Industry 6.0 and additive manufacturing

Industry 6.0 with additive manufacturing demonstrates the potential to transform multiple facets of production processes in the future. The future of Industry 6.0 will introduce enhancements in flexibility and customization, sustainability gains, and intensified manufacturing collaboration [61].

4.5.1.1 Smart, autonomous manufacturing ecosystems

Industries using the advanced technology of Industry 6.0 can implement autonomous manufacturing systems. Upcoming smart factories will establish autonomous production lines by integrating highly advanced sensors with artificial intelligence algorithms and additive manufacturing principles [62]. Independent adaptive production platforms will alter operations according to supply chain network variations while monitoring market requirements and output specifications. Nonhuman-driven predictive analytics paired with IoT technology help manufacturers learn to adjust production schedules automatically, implement fully automated quality management, and handle production supplies. Through AM, manufacturers can produce customized deliveries on demand while minimizing human participation in the entire process. Through manufacturing systems connected to real-time market data, manufacturers can build whole products and customized spare parts instantly while making automatic production changes in response to evolving market trends [63].

4.5.1.2 Personalized consumer products

The advancement of AM technology will make product personalization possible at a historically unattainable level. The fashion industry, automotive sectors, healthcare, and consumer goods production benefit from AM systems, which create personalized products immediately available to match customer requirements [64]. Customers in the fashion field could design items in-store or at local facilities before they are 3D printed for the final products. AM technology will allow healthcare professionals to create personalized medical devices, implants, and prosthetics through precise manufacturing techniques that enhance treatment benefits and patient experience. Advanced manufacturing capabilities founded on AI and Big Data but built using AM science will enable producers to fabricate exact customer-tailored products even through industries formerly dominated by mass-scale production [65].

4.5.1.3 Sustainability and circular economy

The primary trend during Industry 6.0 focuses on sustainability, among other developments. The progress toward sustainable manufacturing will advance through AM technology, which decreases material waste exposure, boosts resource utilization, and prolongs product utilization periods [66]. The circular economy model receives support from AM technologies, which promote material recycling and enable the reuse of resources instead of using fresh materials. The printer technology of the future will use AM to transform recycled materials into products and components that support zero-waste production processes. Industry 6.0 can minimize transportation costs and reduce supply chain emissions and their resulting carbon footprint through localized factory production and on-demand manufacturing [67].

4.5.1.4 Decentralized and localized production

The Industry 6.0 graduation with AM technology increases the possibility of decentralized production. A shift occurs from massive centralized production sites to numerous small, localized manufacturing bases that handle all production duties. Small local printing hubs that meet local manufacturing needs would diminish the requirement for extensive supply chain networks and the need to keep large product inventory stockpiles [68]. This arrangement minimizes transportation and logistics expenses while establishing adaptable manufacturing systems that promptly respond to local market changes. Businesses employing localized production will reduce product delivery times, improving their effectiveness when facing market changes and disruptions such as supply chain delays.

4.5.1.5 Integration with emerging technologies

Industry 6.0 will expand its current integration activities into future technologies, including quantum computing, blockchain networks, and 5G connectivity. The rapid data processing capabilities of quantum computers combined with optimization features allow designers to complete complex additive manufacturing operations hundreds or thousands of times more quickly than possible with traditional systems [69]. Blockchain technology creates a safe digital supply chain management platform with transparent, reliable, and impenetrable material and part-tracking features. Incorporating 5G connectivity, which provides faster and more reliable machine-to-machine communication through network connections, will further increase the efficiency of smart factories and decentralized production systems [70].

4.5.2 Challenges to overcome

Additive manufacturing and Industry 6.0 still have a lot of unexplored potential, but numerous obstacles remain. Manufacturing industries face several challenges, including adopting new technology, integrating systems, adjusting the makeup of their workforce, and managing data security constraints and cost structures.

4.5.2.1 Technological and standardization challenges

Complex technical issues arise when integrating several technologies, such as AI, IoT, AM, and big data. System interoperability is crucial to modern industrial success, but many platforms still use proprietary technologies that are difficult to integrate. Standardization of data transfer between systems is necessary for manufacturing various platforms to manage effective information transfer [71]. Scalability is an issue, mainly when using AM techniques for large-scale manufacturing processes. AM technology is advantageous for small batch production with customization. However, it has drawbacks compared to traditional manufacturing techniques for creating large quantities of goods. Continued research into better materials, faster printing speeds, and enhanced multi-material printing technologies will be necessary to overcome these AM problems [72].

4.5.2.2 Quality control and consistency

A persistent key difficulty that predominantly impacts industries like aerospace, automotive, and healthcare because of their stringent safety rules is quality control and consistency of components using additive manufacturing [73]. Because achieving acceptable strength, precision, and durability standards is difficult, AM technology poses considerable problems for creating intricate, customized components. Variations in materials, printing process abnormalities, and post-fabrication technique issues all contribute to quality issues in additive manufacturing parts. While continuous research and development of specific standards is crucial for industrial components, advanced sensors combined with real-time monitoring systems and AI-based quality control tools will address the current problems in AM production techniques [74].

4.5.2.3 Workforce transformation and skill gaps

The shift to Industry 6.0 necessitates overhauling the conventional staffing strategy. Despite shifts in traditional employment positions caused by AI and robotic automation, computer-supported manufacturing systems will demand more highly skilled workers who supervise, maintain, and run their sophisticated interconnected systems [75]. Workers require data analytics, robotics, artificial intelligence, and advanced manufacturing training to adapt to Industry 6.0 technologies. As robotics evolve, new models of human-machine interactions will be essential, emphasizing the need for workers to develop problem-solving skills.

4.6 CONCLUSION

Additive Manufacturing, together with Industry 6.0, revolutionizes industrial manufacturing through features that include advanced customization capabilities along with improved flexibility and green manufacturing practices. These technologies combine artificial intelligence with the Internet of Things and data analytics methods to build interconnected manufacturing systems that streamline production processes and enable detailed, customized output creation in real time. These paradigm changes allow healthcare industries and automotive businesses to create personalized parts and equipment faster and more efficiently. Industry 6.0 and AM reduce material waste and energy usage by up to 97% compared to traditional manufacturing methods. AM's innovative recycling and dispersed production help minimize transportation-related pollution and support the circular economy. With smart factories and real-time data analysis, machine learning may boost operational efficiency and reduce equipment downtime. However, system integration, quality management, and skilled labor issues hinder the use of these technologies. 3D printing manufacturers must handle material inconsistency while meeting regulatory criteria, especially high-risk items. Properly trained and experienced workers must operate and manage automated systems effectively. Due to high initial investment expenditures, SMEs struggle to integrate these developments.

REFERENCES

1. M. Banafaa et al., "6G mobile communication technology: Requirements, targets, applications, challenges, advantages, and opportunities," *Alexandria Eng. J.*, vol. 64, pp. 245–274, 2023.
2. U. Eswaran, V. Eswaran, V. Eswaran, and K. Murali, "Empowering the Factory of the Future: Integrating Artificial Intelligence, Machine Learning, and IoT Innovations in Industry 6.0," in *Evolution and Advances in Computing Technologies for Industry 6.0*. Boca Raton, FL: CRC Press, pp. 1–21.
3. S. Das and T. Pan, "A strategic outline of Industry 6.0: Exploring the future," *Available SSRN 4104696*, 2022.
4. S. Yadav, S. Rab, and M. Wan, "Metrology and Sustainability in Industry 6.0: Navigating a New Paradigm," in A. Bhatnagar, S. Yadav, V. Achanta, U. Harmes-Liedtke, and S. Rab (eds.), *Handbook of Quality System, Accreditation and Conformity Assessment*. Singapore: Springer, 2023, pp. 1–31.
5. W. Shafik, "Artificial intelligence and internet of things roles in sustainable next-generation manufacturing: An overview of emerging trends in Industry 6.0," in *Sustainable Innovation in Industry 6.0*. Boca Raton, FL: CRC Press/ Taylor & Francis, 2024, pp. 207–239.
6. P. Zahariev et al., "Industry 4.0 and Beyond—Present Trends, Emerging Solutions and Future Technologies in the Area of the Industrial Automation," in *2024 Joint International Conference on Digital Arts, Media and Technology with ECTI Northern Section Conference on Electrical, Electronics, Computer and Telecommunications Engineering (ECTI DAMT & NCON)*, IEEE, January 31–February 3, 2024, Pattaya, Thailand. Piscataway, NJ: Institute of Electrical and Electronics Engineers (IEEE), pp. 525–528.

7. I. Semenets-Orlova, R. Shevchuk, B. Plish, A. Moshnin, Y. Chmyr, and R. Poliuliakh, "Human-centered approach in new development tendencies of value-oriented public administration: Potential of education," *Econ. Aff.*, vol. 65, no. 5, pp. 899–904, 2022.
8. S. Kumar et al., "Machine learning techniques in additive manufacturing: a state of the art review on design, processes and production control," *J. Intell. Manuf.*, vol. 34, no. 1, pp. 21–55, 2023.
9. S. Sahoo and C.-Y. Lo, "Smart manufacturing powered by recent technological advancements: A review," *J. Manuf. Syst.*, vol. 64, pp. 236–250, 2022.
10. S. Chourasia, S. M. Pandey, and A. K. Keshri, "Prospects and challenges with legal informatics and legal metrology framework in the context of industry 6.0," *MAPAN*, vol. 38, no. 4, pp. 1027–1052, 2023.
11. M. R. Nair, N. Bindu, R. Jose, and K. Satheesh Kumar, "From assistive technology to the backbone: The impact of blockchain in manufacturing," *Evol. Intell.*, vol. 17, no. 3, pp. 1257–1278, 2024.
12. N. Ryvak, "Znaczenie cyfryzacji i innowacji w ramach rewolucji przemysłowej w kontekście współczesnych wyzwań," *Zesz. Nauk. Wyższej Szk. Bank. w Pozn.*, vol. 100, no. 1, pp. 95–108, 2023.
13. R. B. Nambiar, C. P. Dheeshith, K. Abhijith, A. Shahaas, and R. R. Gatti, "Implementing Cyber-Physical Systems in Manufacturing Systems," in *Emerging Technologies and Digital Transformation in the Manufacturing Industry*. Hershey, PA: IGI Global, 2023, pp. 150–172.
14. H. Pack and L. E. Westphal, "Industrial strategy and technological change: theory versus reality," *J. Dev. Econ.*, vol. 22, no. 1, pp. 87–128, 1986.
15. S. Kumar, G. Tiwari, and N. Tiwari, "Evolving Industries: A Journey from Industry 1.0 to 5.0," in *Artificial Intelligence and Communication Techniques in Industry 5.0*. Boca Raton, FL: CRC Press, pp. 73–86.
16. P. Subbiah, A. K. Tyagi, and B. D. Mazumdar, "The Future of Manufacturing and Artificial Intelligence Industry 6.0 and Beyond," in *Industry 4.0, Smart Manufacturing, and Industrial Engineering*. Boca Raton, FL: CRC Press, 2024, pp. 347–362.
17. J. Mokyr, "Editor's Introduction: The New Economic History and the Industrial Revolution," in *The British Industrial Revolution: An Economic Perspective* (5th ed.). London: Routledge, 2018, pp. 1–127.
18. A. J. Barbour, *Mechanical Powers: Engineering and Romantic Poetics in the Early Anthropocene*. University of California, Berkeley, 2021.
19. A. S. George, "The fourth industrial revolution: A primer on Industry 4.0 and its transformative impact," *Partners Univers. Innov. Res. Publ.*, vol. 2, no. 1, pp. 16–40, 2024.
20. M. Javaid, A. Haleem, R. P. Singh, and R. Suman, "Substantial capabilities of robotics in enhancing industry 4.0 implementation," *Cogn. Robot.*, vol. 1, pp. 58–75, 2021.
21. D. G. S. Pivoto, L. F. F. De Almeida, R. da Rosa Righi, J. J. P. C. Rodrigues, A. B. Lugli, and A. M. Alberti, "Cyber-physical systems architectures for industrial internet of things applications in industry 4.0: A literature review," *J. Manuf. Syst.*, vol. 58, pp. 176–192, 2021.
22. C. Li, P. Zheng, Y. Yin, B. Wang, and L. Wang, "Deep reinforcement learning in smart manufacturing: A review and prospects," *CIRP J. Manuf. Sci. Technol.*, vol. 40, pp. 75–101, 2023.

23. N. S. Kolhe, A. Panicker, S. Mhatre, and J. Vichare, "Industry 4.0 and chemical industry: The technology enablers behind strategic growth of chemical companies," *Int. J. Res. Appl. Sci. Eng. Technol.*, vol. 11, no. 1, pp. 58–71, 2023.
24. R. Rame, P. Purwanto, and S. Sudarno, "Industry 5.0 and sustainability: An overview of emerging trends and challenges for a green future," *Innov. Green Dev.*, vol. 3, no. 4, p. 100173, 2024.
25. A. Almusaed, I. Yitmen, and A. Almssad, "Reviewing and integrating aec practices into industry 6.0: Strategies for smart and sustainable future-built environments," *Sustainability*, vol. 15, no. 18, p. 13464, 2023.
26. S. Pattanaik, M. Mohammed, and V. Sood, "Artificial Intelligence and Machine Learning in Industry 6.0," in *Industry 6.0*. Boca Raton, FL: CRC Press, 2024, pp. 37–54.
27. C. Vezzoli, L. Macrì, B. Takacs, and D. Yang, *System Design for Sustainability in Practice. Methods, Tools and Guidelines to Design Sustainable Product-Service Systems Applied to Distributed Economies*. Milan: Politecnica, 2022, pp. 1–276.
28. H. ElMaraghy, L. Monostori, G. Schuh, and W. ElMaraghy, "Evolution and future of manufacturing systems," *CIRP Ann.*, vol. 70, no. 2, pp. 635–658, 2021.
29. S. Sharma, S. S. Joshi, M. V. Pantawane, M. Radhakrishnan, S. Mazumder, and N. B. Dahotre, "Multiphysics multi-scale computational framework for linking process–structure–property relationships in metal additive manufacturing: A critical review," *Int. Mater. Rev.*, vol. 68, no. 7, pp. 943–1009, 2023.
30. O. Abdulhameed, A. Al-Ahmari, W. Ameen, and S. H. Mian, "Additive manufacturing: Challenges, trends, and applications," *Adv. Mech. Eng.*, vol. 11, no. 2, p. 1687814018822880, 2019.
31. S. C. Vetrivel, T. Gomathi, K. C. Sowmiya, and V. Sabareeshwari, "Customer-Centric Excellence in the Marketing 6.0 Era: Industry 6.0," in *Advanced Businesses in Industry 6.0*. Hershey, PA: IGI Global, 2024, pp. 192–219.
32. Y. Huang, M. C. Leu, J. Mazumder, and A. Donmez, "Additive manufacturing: Current state, future potential, gaps and needs, and recommendations," *J. Manuf. Sci. Eng.*, vol. 137, no. 1, p. 14001, 2015.
33. O. Chenaru, S. Mocanu, R. Dobrescu, and M. Nicolae, "Enhancing antifragile performance of manufacturing systems through predictive maintenance," *Appl. Sci.*, vol. 12, no. 23, p. 11958, 2022.
34. L. J. Ladani, "Applications of artificial intelligence and machine learning in metal additive manufacturing," *J. Phys. Mater.*, vol. 4, no. 4, p. 42009, 2021.
35. Z. M. Çınar, A. Abdussalam Nuhu, Q. Zeeshan, O. Korhan, M. Asmael, and B. Safaei, "Machine learning in predictive maintenance towards sustainable smart manufacturing in industry 4.0," *Sustainability*, vol. 12, no. 19, p. 8211, 2020.
36. M. Soori, B. Arezoo, and R. Dastres, "Internet of things for smart factories in industry 4.0, a review," *Internet Things Cyber-Phys. Syst.*, vol. 3, pp. 192–204, 2023.
37. Y. Zhang, S. Ren, Y. Liu, and S. Si, "A big data analytics architecture for cleaner manufacturing and maintenance processes of complex products," *J. Clean. Prod.*, vol. 142, pp. 626–641, 2017.

38. Y.-M. Tang, G. T. S. Ho, Y.-Y. Lau, and S.-Y. Tsui, "Integrated smart warehouse and manufacturing management with demand forecasting in small-scale cyclical industries," *Machines*, vol. 10, no. 6, p. 472, 2022.
39. M. Doyle-Kent and P. Kopacek, "Optimising Human Potential Through Diversity and Inclusion for Industry/Production 4.0, 5.0 and 6.0," in *The International Symposium for Production Research*. Cham: Springer, 2022, pp. 267–276.
40. M. A. Musarat, M. Irfan, W. S. Alaloul, A. Maqsoom, and M. Ghufran, "A review on the way forward in construction through industrial revolution 5.0," *Sustainability*, vol. 15, no. 18, p. 13862, 2023.
41. K. Kanishka and B. Acherjee, "Revolutionizing manufacturing: A comprehensive overview of additive manufacturing processes, materials, developments, and challenges," *J. Manuf. Process.*, vol. 107, pp. 574–619, 2023.
42. M. P. Groover, *Automation, Production Systems, and Computer-Integrated Manufacturing* (4th ed.). Noida: Pearson Education India, 2016.
43. E. Annanperä et al., "From industry x to industry 6.0: antifragile manufacturing for people, planet, and profit with passion," in *Proceedings of the Design Society*, vol. 1. Cambridge: Design Society. 2021, pp. 2863–2872.
44. S. Das, A. Chowdhury, and R. Sil, "Quantum-Powered Industry 6.0 Innovations for Enhanced Catastrophe Management," in *The Rise of Quantum Computing in Industry 6.0 Towards Sustainability*. Cham: Springer, 2024, pp. 67–79.
45. A. Gonçalves, B. Ferreira, M. Leite, and I. Ribeiro, "Environmental and economic sustainability impacts of metal additive manufacturing: A study in the industrial machinery and aeronautical sectors," *Sustain. Prod. Consum.*, vol. 42, pp. 292–308, 2023.
46. A. S. Gohardani, G. Doulgeris, and R. Singh, "Challenges of future aircraft propulsion: A review of distributed propulsion technology and its potential application for the all electric commercial aircraft," *Prog. Aerosp. Sci.*, vol. 47, no. 5, pp. 369–391, 2011.
47. J. Heilala and J. Kantola, "Sustainable Manufacturer Engineering for Industry 6.0," in *International Conference on Axiomatic Design*. Cham: Springer, 2023, pp. 75–87.
48. M. M. Roshid, A. Waaje, and R. Karim, "Industry 6.0 as an emerging research field: A systematic and bibliometric analysis", in *Proceedings of the EMIDWORLD 2nd International Congress on Economics, Public Finance, Business & Social Sciences*, November 7–9, 2024, Istanbul, Türkiye, pp. 1–24.
49. A. Keshari and P. Singh, "Analysis of Industry 4.0 and Industry 5.0, investigating their evolution and collaboration," *Interdiscip. J. Contemp. Res*, vol. 10, pp. 21–28, 2023.
50. M. Erdoğan, "Strategic Ranking for Industry 6.0 Transition: Applying Fuzzy Hybrid Multi-Criteria Analysis in Manufacturing," *Eng. Technol. Appl. Sci. Res.*, vol. 14, no. 5, pp. 16868–16877, 2024.
51. A. K. Tyagi, P. Bhatt, N. Chidambaram, and S. Kumari, "Artificial intelligence empowered smart manufacturing for modern society: A review," in *Artificial Intelligence, Digital Twin, and Smart Manufacturing*. Boca Raton, FL: CRC Press, 2024, pp. 55–83.
52. R. Jain, "Towards a Green Revolution: Sustainable Integration of Industry 6.0 Technologies and Smart Banking Services in the UAE," in *Sustainable Innovation for Industry 6.0*. Hershey, PA: IGI Global, 2024, pp. 31–44.

53. J. V. Kumar, K. Radhakrishnan, and L. S. Priya, "Shaping Tomorrow: The Impact of 3D Printing on Functional Material Fabrication," in *Breaking Boundaries: Pioneering Sustainable Solutions Through Materials and Technology*. Cham: Springer, 2025, pp. 281–306.
54. D. Strong, M. Kay, B. Conner, T. Wakefield, and G. Manogharan, "Hybrid manufacturing–integrating traditional manufacturers with additive manufacturing (AM) supply chain," *Addit. Manuf.*, vol. 21, pp. 159–173, 2018.
55. S. Cecchel, "Materials and technologies for lightweighting of structural parts for automotive applications," *SAE Int. J. Mater. Manuf.*, vol. 14, no. 1, pp. 81–98, 2021.
56. J. Vijay Arputharaj, B. N. J. William, A. A. Haruna, and D. D. Prasad, "Exploring the synergy of IIoT, AI, and data analytics in industry 6.0," in *Industry 6.0: Technologies, Practices, Challenges, and Applications*, vol. 1. Boca Raton, FL: CRC Press, 2024, pp. 85–108.
57. S. Kumari and A. K. Tyagi, "Online social networking: Power of Industry 6.0 and beyond," in *Online Social Networks and Business Frameworks*. Boca Raton, FL: CRC Press, 2024, pp. 255–298.
58. A. S. Duggal et al., "A sequential roadmap to industry 6.0: Exploring future manufacturing trends," *IET Commun.*, vol. 16, no. 5, pp. 521–531, 2022.
59. U. Punia et al., "Adopting additive manufacturing: Spotlight on sustainability and material utilization," *Green Mater.*, vol. 13, pp. 1–34, 2025.
60. M. Chahar, R. Kumar, A. Habeeb, R. Kumar Garg, and U. Punia, "Factors affecting flexural and impact strength of natural fiber reinforced polymer composites: A review," *Green Mater.*, pp. 1–40, Sep. 2024, doi: 10.1680/jgrma.24.00108.
61. A. Tyagi, U. Punia, A. Dadhich, and S. L. Meena, "Global perspective of natural fibre reinforced composites: Properties, and applications," *J. Inst. Eng. Ser. C*, pp. 1–16, 2024.
62. U. Punia, A. Kaushik, R. K. Garg, A. Sharma, and D. Chhabra, "Additive Manufacturing Requirements for Dental Implementation," in *3D Printing and Sustainable Product Development*. Boca Raton, FL: CRC Press, 2023, pp. 151–171.
63. A. Kaushik, U. Punia, S. Gahletia, R. K. Garg, and D. Chhabra, "Identification and Overcoming Key Challenges in the 3D Printing Revolution," in *3D Printing and Sustainable Product Development*. Boca Raton, FL: CRC Press, 2023, pp. 87–104.
64. U. Punia and R. K. Garg, "Reinforcing dental restoration material PMMA with SS 316L nanoparticles for mechanical performance analysis," in *Progress in Additive Manufacturing*. Cham: Springer, 2024, pp. 1–12.
65. U. Punia and R. K. Garg, "Enhancing mechanical performance of 3D printable PMMA resin through strategic incorporation of SS 316 L nanoparticles for dental applications," *Int. J. Interact. Des. Manuf.*, vol. 18, no. 8, pp. 6317–6332, 2024.
66. C. Gao, S. Wolff, and S. Wang, "Eco-friendly additive manufacturing of metals: Energy efficiency and life cycle analysis," *J. Manuf. Syst.*, vol. 60, pp. 459–472, 2021.
67. R. Nanthan and K. Pitchaimani, "Enhancing Resilience and Sustainability in the Automotive Supply Chain with Additive Manufacturing: A Simulation-Driven Virtual Twin Strategy," *J. Ind. Inf. Integr.*, vol. 45, Article 100712.

68. S. H. Khajavi, J. Holmström, and J. Partanen, "Additive manufacturing in the spare parts supply chain: Hub configuration and technology maturity," *Rapid Prototyp. J.*, vol. 24, no. 7, pp. 1178–1192, 2018.
69. C. Tian et al., "Data-driven approaches toward smarter additive manufacturing," *Adv. Intell. Syst.*, vol. 3, no. 12, p. 2100014, 2021.
70. T. M. Fernandez-Carames and P. Fraga-Lamas, "A review on the application of blockchain to the next generation of cybersecure industry 4.0 smart factories," *IEEE Access*, vol. 7, pp. 45201–45218, 2019.
71. T. P. Raptis, A. Passarella, and M. Conti, "Data management in industry 4.0: State of the art and open challenges," *IEEE Access*, vol. 7, pp. 97052–97093, 2019.
72. A. Nazir et al., "Multi-material additive manufacturing: A systematic review of design, properties, applications, challenges, and 3D printing of materials and cellular metamaterials," *Mater. Des.*, vol. 226, p. 111661, 2023.
73. L. Zhou et al., "Additive manufacturing: A comprehensive review," *Sensors*, vol. 24, no. 9, p. 2668, 2024.
74. Z. Bin Akhtar, "Artificial intelligence (AI) within manufacturing: An investigative exploration for opportunities, challenges, future directions," *Metaverse*, vol. 5, no. 2, p. 2731, 2024.
75. D. Stadnicka et al., "Industrial needs in the fields of artificial intelligence, internet of things and edge computing," *Sensors*, vol. 22, no. 12, p. 4501, 2022.

Chapter 5

Advancing environmental sustainability with functionally graded biomaterials in additive manufacturing

Deepak Gupta, Arun Kumar Chaudhary, Alka Verma, Dipesh Kumar Mishra, Vikas Misra, and Shreejaya Sivadas

5.1 INTRODUCTION

Functionally graded biomaterials represent a new frontier of material science for a variety of applications, particularly in biomedicine. These materials are designed with a gradual or step change in their composition and structure, which allows for unique properties, each particularly suitable for biological and mechanical demands (Goto, 2015). As a result, FGBMs with AM have been developed as a novel strategy to address several challenges facing traditional biomaterials, including poor mechanical integration with host tissues and mismatch in properties (Goto, 2018).

Use of materials AM has opened a world of possibilities in terms of material use, providing precision in fabrication while minimizing waste. AM differs from traditional subtractive manufacturing techniques by minimizing material waste, especially in fabricating functionally graded biomaterials (FGBMs). The layer-by-layer deposition enables only the amount of material required to be used, making AM a facilitator of sustainable manufacturing practices.

The concept of FGBMs with AM is entirely inspired by natural systems, which have a gradient in the material properties. For instance, hard and dense outer cortical bone evolves into a spongy and porous trabecular bone towards the middle, implying that bone has a gradient in composition and structure. This optimizes both the bone's mechanical and biological functionality (Emir & Bahçe, 2023). For medical applications, FGBMs with AM seek to replicate this to offer better performance. Biocompatibility and biodegradation are two major aspects of FGBM related to biomedical applications. Biocompatibility is considered the aptitude of a material to perform well in the body with an appropriate host response when implanted. A host response includes no immune reactions, toxicity, and inflammation. Biocompatibility should be high for long-term implant and other biomaterial durability and function (Enab, 2012).

In contrast, biodegradation is the gradual degradation of materials by biological systems. Biodegradation control is crucial for tissue engineering and regenerative medicine. Degradation should allow tissue

DOI: 10.1201/9781003607052-5

reinstatement and decrease throughout the time framework (Watanabe et al., 2004). This Biodegradation is necessary to avoid future issues with the material. This chapter explains the fundamental biocompatibility and biodegradation characteristics of FGBM. That is, we present basic design principles, interaction mechanisms, and their applications. The key goal is to give each professional insight into how FGBM can be modified to better fit the unique clinical settings within each profession and ultimately enhance the patient's clinical outcomes. The difficulty and novelty of biomedical material needs mean that well-understood FGBM biocompatibility and biodegradation needs will have to be addressed through cross-disciplinary approach (Hedia, Shabara, et al., 2006). Indeed, such an understanding should include standard arrangements of material science and technology, biotechnology, and medical measures in the future. Such integration into new FGBM can improve not only the structural and mechanical aspects of the material but also the characteristics of its connections with human or animal tissue. There is even more promise for FGBM application to form new medical applications as the technology progresses, which would benefit numerous potential patients.

Table 5.1 highlights the specific advantages of integrating functionally graded biomaterials with additive manufacturing, providing a quick overview of their contributions to biomedicine and sustainability.

Table 5.1 Advantages of FGBMs with AM for additive manufacturing

Advantage	*Details*
Material efficiency	Reduces material waste with precise layer-by-layer deposition.
Customization	Enables patient-specific designs for implants and scaffolds.
Biocompatibility	Enhances tissue integration and reduces adverse immune responses.
Controlled degradation	Supports tissue regeneration and avoids foreign body complications.
Gradual property transition	Mimics natural tissue gradients for optimized functionality.
Lightweight structures	Ideal for applications requiring minimal material usage.
Energy efficiency	Lowers manufacturing energy consumption compared to traditional methods.
Versatile material compositions	Allows integration of ceramics, polymers, and composites.
Enhanced mechanical performance	Provides tailored strength and flexibility for various applications.
Circular economy compatibility	Promotes recyclability and biodegradability for sustainability.

5.2 BIOCOMPATIBILITY OF FUNCTIONALLY GRADED BIOMATERIALS

Biocompatibility is a pivotal quality for biomaterials, characterized by a material's aptitude to carry out appropriately with a fitting host response in a targeted application. Within the setting of functionally graded biomaterials, biocompatibility signifies that the substance must not provoke an adverse immune response, toxicity, or swelling when implanted in the body. The critical nature of biocompatibility cannot be overemphasized, as it determines the success and longevity of surgical implants and instruments (Afzal et al., 2012). A biocompatible material ensures that the body acknowledges the implantation, incorporates it with the surrounding tissues, and maintains its functionality through time without inducing harm to the patient.

Functionally graded biomaterials are designed to mimic the natural gradients found in living tissues, such as bone, which exhibit variation in composition and properties to optimize mechanical strength and biological functionality. By achieving a high level of biocompatibility, functionally graded biomaterials can furnish enhanced execution in numerous biomedical applications (Figure 5.1), from bone and dental implants to tissue engineering scaffolds and drug conveyance frameworks (Leong et al., 2008).

5.2.1 Factors influencing biocompatibility

There are many factors that come into play when discussing the biocompatibility of FGBMs with AM. These include surface, mechanical properties, as well as the chemical and gradual composition of the material. These factors must be carefully understood and optimized to design the material suitable for incorporation (Figure 5.2) and effective operation in the human body (Madeira et al., 2018).

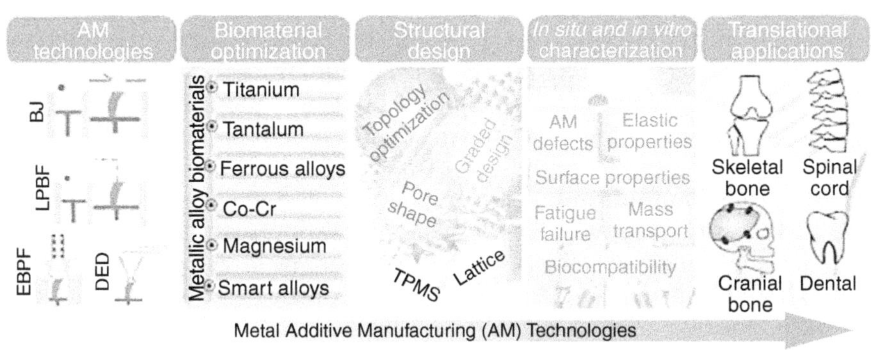

Figure 5.1 Metal additive manufacturing (AM) technologies (Davoodi et al., 2022).

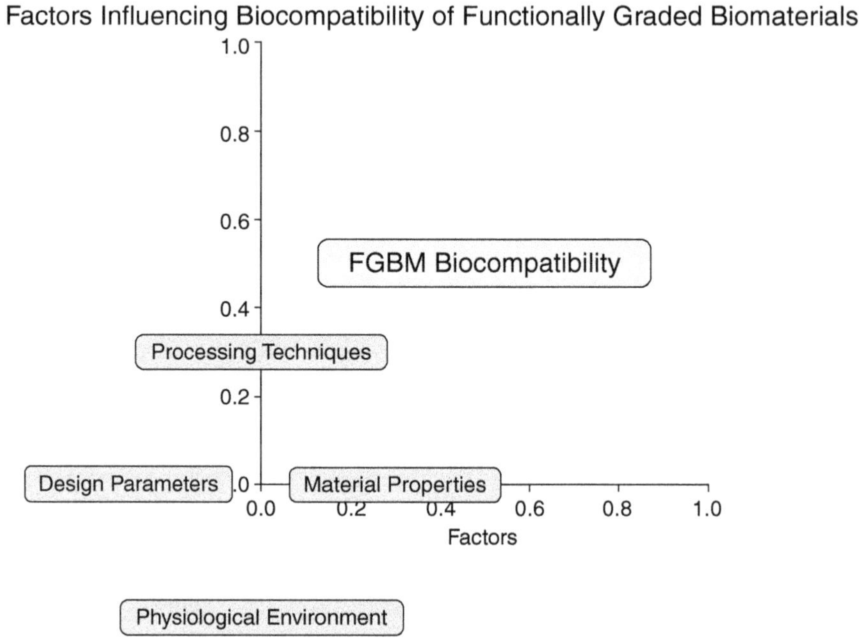

Figure 5.2 A multifaceted approach: factors influencing FGBM biocompatibility.

As shown by the mind map in Figure 5.2, the vast number of variables make this issue a complex one. This implies that a holistic approach must be taken when designing and developing FGBMs with AM.

Table 5.2 consolidates the key factors affecting biocompatibility, providing a quick reference for readers to understand how these aspects interact in biomedical applications.

5.2.1.1 Surface chemistry

Surface chemistry is critical for biocompatibility as it forms the interface between biomaterials and living tissues. The outermost layer of an implant interacts initially with the body, so its chemical and physical attributes dictate how surrounding cells and systems will respond.

- **Surface modification:** This interaction can be further optimized through surface modifications. Incorporating materials with signaling molecules, proteins, or growth factors to direct cell attachment, growth, and specialization on the material. Coatings with a surface composition mimicking that of a bone, such as hydroxyapatite or bioactive glass, are most used to improve bone graft integration (Lin et al., 2010).

Table 5.2 Key factors influencing biocompatibility of FGBMs in additive manufacturing

Factor	*Details*
Surface chemistry	Interactions between material surface and biological tissues.
Surface modification	Addition of signaling molecules or proteins to enhance cell adhesion.
Protein adsorption	Attracting proteins to facilitate cell integration.
Surface roughness	Micro/nano-scale texturing for improved tissue attachment.
Chemical composition	Presence of functional groups aiding cell-surface interaction.
Mechanical properties	Match between biomaterial and host tissue mechanics.
Stress distribution	Reduces stress shielding, ensuring better tissue integration.
Fatigue resistance	Prolonged implant lifespan under physiological loads.
Gradual composition	Smooth material transitions minimize interface problems.

- **Protein adsorption:** When implanted, a material's surface quickly becomes covered with adsorbed proteins from the surrounding fluids in the environment. Then these adsorbed proteins' nature and conformation regulate the characteristics and behaviors of the latter reacting cells. They can enhance tissue integration by attracting proteins that support adhesion and ingrowth of cells (Kabiraj et al., 2023).
- **Surface roughness and topography:** Physical topography also strongly impacts biocompatibility. Micro- and nano-scale surface texturing can steer cell attachment, orientation, and behavior through topographical cues. For example, a roughened external layer may facilitate improved initial cell binding compared to a smooth one, permitting stronger tissue incorporation (Geoghegan et al., 2023).
- **Chemical composition:** Elemental and molecular structure at the interface is another determinant. Functional groups like hydroxyl, carboxyl, or amine groups which permit robust interactions with cellular receptors can augment compatibility by facilitating cell-surface interactions (Tirta Nindhia et al., 2008).

5.2.1.2 Mechanical properties

The mechanical performance of functionally graded biomaterials plays an indispensable role in determining their biocompatibility and overall performance in biomedicine. FGBMs with AM' distinctive feature is the ability to vary in their mechanical behavior that can be achieved by having the gradient for each of the abovementioned mechanical properties (Prakash & Singh, 2020).

- **Tissue properties:** The primary aim of the FGBMs with AM design is to match the mechanical properties of tissues with which these biomaterials will interact. For instance, bone has a substantial difference in its mechanical properties from hard, dense cortical bone to soft, porous trabecular bone. A decreased stiffness gradient could reduce stress shielding and subsequent bone resorption around the implant (Scaffaro et al., 2016).

- **Stress distribution:** A diverse material composition could promote the transfer of stress from implant to the tissue. This allows substantial robust mechanical performance that shifts the stress load to tissue. Decreased stiffness gradients help minimizing stress-shielding and hence, reduce tissue resorption.
- **Fatigue resistance:** Fatigue life is paramount for extended implant occupation as evidenced by professional practice. For example, the expansion of an implant lifespan aligns with clinical professional life's success by promoting implant usage. The available option allows for increasing several materials to varying the component, with allowance to control biodegradability (Xiong et al., 2013).
- **Tailor mechanical properties:** Applications such as dental implants need to be rigid and withstand the resultant shaping force; in contrast, vascular graft such as heart stents need to remain fluid, and this gradient would provide this compatibility (Veloso et al., 2022).

5.2.1.3 Gradual composition

This gradual concept in FGBMs with AM is defining and sets them apart from traditional homogeneous biomaterials. This gradient composition allows a smooth transition between the different phases of material, enhancing biocompatibility as well as improving functionality.

- **Optimized biological integration:** Since the gradual composition can be used to optimize biological integration at different locations, this feature can be designed to match. For example, in a bone implant, the outer layer may be designed to be highly osteoconductive to promote bone ingrowth, while the inner epilayers are for mechanical strength, enhancing the performance of the implant (Zhu et al., 2018).
- **Minimized interface problems:** Conventional implant materials are known to have interface problems such as delamination and poor metal- to- tissue adhesion. The composition gradient technique of FGBMs with AM helps in minimizing these problems, since a gradual transition ensures that there is no material remaining as a foreign body (Hedia, El-Midany, et al., 2006). When FGBM is well integrated with the surrounding tissue, it is very stable and also does not elicit an inflammatory response.
- **Controlled degradation rates:** In biodegradable FGBMs with AM, the varying composition gradient between different regions can be utilized to control the degradation rate. This is specifically beneficial in tissue engineering whereby the outer layers are needed to degrade rapidly to

allow for the cell attachment and proliferation. The inner layers are to be preserved longer to provide support as the tissues regenerate. This ensures that the scaffold degrades in harmony with the tissue healing (Y. Li et al., 2019).

- **Functional gradients:** Other than mechanical and biological properties, the various composition gradient can be used to provide other gradients such as thermal or electrical properties (Mohammadi et al., 2021).

5.2.2 Biocompatibility testing

Ensuring the biocompatibility of functionally graded biomaterials is essential for the safe application of these materials in the biomedical field. In general, biocompatibility is tested using in vitro and in vivo methods, in which various factors such as toxic effects, cell adhesion, proliferation, and tissue response are evaluated.

5.2.2.1 *In vitro tests*

In vitro tests are laboratory-based experiments conducted outside of living organisms. They are usually preliminary measures to assess FGBMs with AM' interactions with living body systems and are aimed at helping screen the materials to be further assessed under in vivo circumstances (Werner et al., 2002).

5.2.2.1.1 Cytotoxicity assays

The cytotoxicity tests assay the occurrence of cytotoxic substances in the tested material. They include:

- **Purpose:** To evaluate the release of toxic substances capable of killing or damaging cells. These assays are necessary for determining the safety of the biomaterial.
- **Methods:** MTT assay, XTT assay, and LDH release assay are the most common. They measure metabolic activity or cell viability after exposure to the material or its extracts (Bahraminasab et al., 2022).
- **Procedure:** Cells are cultured with the material extracts or in direct contact with the material. Then, incubation is performed, after which cell viability is assessed using colorimetric and fluorometric indicators.

5.2.2.1.2 Cell adhesion and proliferation studies

- **Purpose:** To explore the favorable adhesion and proliferation of cells to the FGBM surface. This validates the possibility of biocompatibility and integration of the implant.

- **Methods:** It is possible to conduct a microscopic assessment, cell quantification and observation, as well as the use of standard fluorescently dyed cells.
- **Procedure:** Creation of sterile cell seeding on the material surface and incubation, followed by the quantification of the number of cells adhered and proliferated with time. The investigation observed for both light microscope and fluorescent substance inclusion (Bambole & Yakhmi, 2016).

5.2.2.1.3 Protein adsorption tests

- **Purpose:** Protein adsorption testing is conducted for the purpose of understanding the material's ability to adsorb proteins dispersed in the adjacent biological environment, which is a key factor in cell-material interaction.
- **Methods:** The quantity of the adsorbed proteins along with their type is determined by the means of enzyme-linked immunosorbent assay and surface plasmon resonance.
- **Procedure:** The material's surface is exposed to the protein solution, after which the adsorbed proteins are quantified and the analysis is performed (Lvov et al., 2016).

5.2.2.1.4 Hemocompatibility tests

- **Purpose:** The hemocompatibility of FGBMs with AM is tested to examine how these materials interact with blood elements. This will help determine whether FGBMs with AM are applicable for cardiovascular biomaterial applications.
- **Methods:** Platelet adhesion assays, hemolysis levels, and coagulations are some of the tests that can be conducted based on the target to measure these parameters.
- **Procedure:** It is done by exposing the material to blood samples by inserting it into the cleaned glass, where platelet adhesion between the glass and the synthesized materials will be determined. Sometimes, the work might need some additional lab equipment or materials to enable accurate measurements (Li et al., 2023).

5.2.2.2 In vivo tests

In vivo tests refer to the implantation of the material into a living being to closely study the general biological response in a continuously changing environment. In vivo tests offer all-rounded information regarding biocompatibility (Shah et al., 2019). The information provided may include an observable immune reaction, integration with tissue, and response to functionality.

5.2.2.2.1 *Animal models*

- **Purpose:** The use of animal models to assess the biocompatibility of FGBMs with AM in a living body is essential to observe the metrics of the material in contact with tissues and organs.
- **Common model:** Models used include rodents ranging from rats, mice, rabbits, and larger animals such as pigs or sheep depending on the use.
- **Practicality:** Surgical implantation of the material in the target tissue in animal models and continued examination of the animal model provides information on inflammation, infection and rejection of the implanted material. Histological analysis of the tissue's cells enhances the integration (Ma et al., 2009).

5.2.2.2.2 *Histological analysis*

- **Purpose:** Examination of histological changes in the cellular response of the tissue surrounding the implanted material. Thus, the material should be stained for proper investigation.
- **Methods:** Obtained tissues are first fixed, and then staining technique such as hematoxylin and eosin are used (Wang et al., 2024). Other staining techniques which are applicable are Immunohistochemistry and electron microscopy.
- **Procedure:** Stained tissue sections are examined by a microscope to evaluate cell infiltration and the regeneration of the tissue, the magnitude of inflammatory cells.

5.2.2.2.3 *Immunological response*

- **Purpose:** Using immunology and immunohistochemistry, the immune response to FGBM was analyzed, enabling us to predict the material's behavior in the body in the long run.
- **Methods:** Blood tests, those for the cytokine assay, and a histological examination were conducted.
- **Procedure:** Blood biochemistry markers and other cytokines were quantified with blood samples. Examination of tissues for immune cells, such as macrophages and lymphocytes (Naahidi et al., 2013).

5.2.2.2.4 *Functional integration*

- **Purpose:** Assessment of the material's functional viability in an application – whether the material can perform as intended across the host tissue.
- **Methods:** The function is usually specific to a particular operation and condition, such as mechanical integrity for bone implants or electrical connectivity for neural interfaces (Xu et al., 2014).

- **Procedure:** Implant function determines the material's performance qualities across physiological conditions that fulfill their function. The conducted analyses determine several micro- or macroscopic structures required for the development of a functional biocompatible material.

Biocompatibility testing of FGBMs with AM is required, starting from in vitro studies to in vivo studies (Bogdanski et al., 2002). This will give the necessary information about how the material acts in the biological systems of the host, thus confirming that they are fit to be used for in situ conditions and biocompatible. By undertaking this extensive biocompatibility testing of these materials, scientists can develop FGBMs with AM that are not just effective but also can be used for several clinical years.

5.3 BIODEGRADATION OF FUNCTIONALLY GRADED BIOMATERIALS

The process by which a living organism breaks down a material into compounds acceptable to the body to be emitted or absorbed. Biodegradation is critical for the appropriate application of functionally graded biomaterials. "Controlled biodegradation" is the term used to describe the phenomenon (Gibson et al., 2002). The application of FGBMs with AM where the biodegradation of newly developing tissue is a primary objective is another prime example. Biodegradation is necessary as it allows temporary support structures and reduces the need for a second surgery to extract implants and foreign body response in the long term. Biodegradation of FGBMs with AM in tissue engineering must be regulated such that it occurs within the rate of formation of new tissue.

5.3.1 Factors affecting biodegradation

The biodegradability of FGBMs allows them to be compatible with biological systems and to mitigate the waste management challenges caused by conventional FGBMs. By incorporating organic materials into additive manufacturing processes, it is possible to produce temporary systems like medical scaffolds, which degrade and do not require disposal and accumulation of waste over time.

There are several factors that influence the biodegradation of FGBMs with AM, including material composition, porosity and structure, and environmental conditions within the body (Figure 5.3).

Understanding and optimizing these factors are important for designing FGBMs with AM with predictable and beneficial degradation profiles.

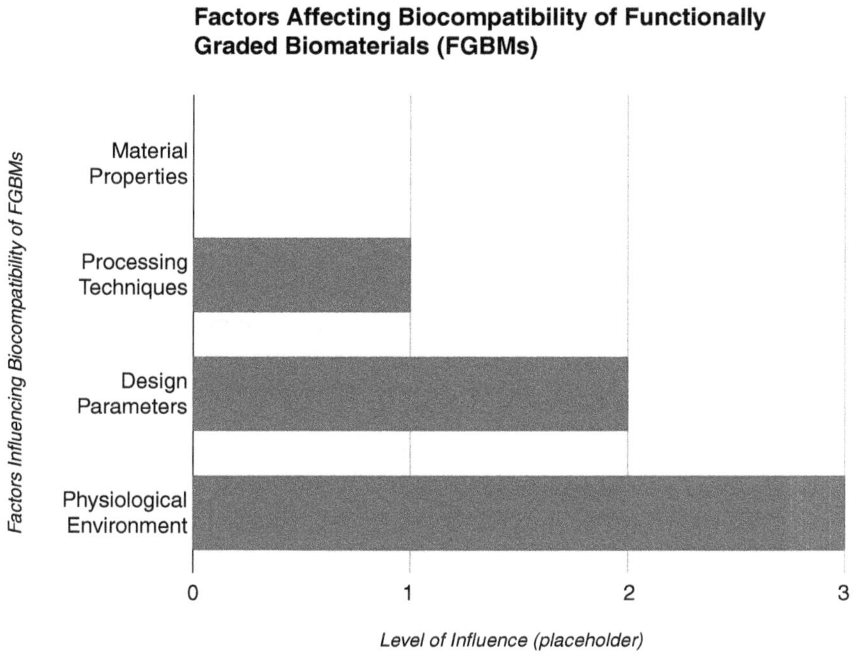

Figure 5.3 Key factors impacting the biocompatibility of functionally graded biomaterials.

5.3.1.1 Material composition

The composition of FGBMs with AM has a significant impact on their degradation. Different materials are degraded under various modes and with different kinetics, including by hydrolysis, enzymatic degradation, or cellular processes.

- **Polymeric materials:** Many FGBMs with AM comprise biodegradable polymers like polylactic acid, polyglycolic acid, and their copolymers, including PLGA. These polymers are degraded by hydrolytic cleavage of their ester bonds, and their rate of degradation is modulated via adjustment of their molecular weight, crystallinity, and copolymer ratio (Qasim et al., 2020).
- **Ceramic materials:** Another type of material is biodegradable ceramics, including calcium phosphates like hydroxyapatite and tricalcium phosphate, which are degraded by solution and osteoclast resorption. Due to the bioactivity of calcium phosphates and their resemblance to bone mineral, they are commonly used in bone regeneration systems (Dorozhkin, 2010).
- **Composite materials:** Additionally, FGBMs with AM often include composite materials, which combine the advantageous properties of

both types of materials. For example, a composite material consisting of PLA and hydroxyapatite can be used to provide structural integrity and bioactivity; its degradation can be regulated by adjustments of the components' content and distribution (Achrai & Wagner, 2017; Gupta et al., 2023; Gupta et al., 2024).

5.3.1.2 Porosity and structure

The degradation rate of FGBMs with AM is highly influenced by their internal structure and porosity. High porosity increases the surface area exposed to biological fluids and cells, increasing the material's degradation kinetics. It is provided that materials with higher porosity incorporate more surfaces for enzymatic and hydrolytic activity, thus degrading faster.

- **High porosity** is beneficial for clinical use in cases of rapid tissue integration or ingrowth. However, it is necessary to standardize this feature, as uncontrolled porosity can lead to rapid material resorption.
- **Controlled porosity** can also be provided through porosity gradients in FGBMs with AM. When degradation rates are prominent, cells are provided with more space for penetration and integration. In vitro technology enables the recreation of a space-filling balance via temperature and the humidity of vapor. Porosity gradients allow changes in tissue properties. For example, the outer layer of those products is typically relatively dense for temporal mechanical support. Eventually, however, the inner portion is more porous than that enabled by the porosity gradient, which will enable cellular infiltration and vascularization. Therefore, porous structure customization is advanced through these techniques.
- **The microstructure** of FGBMs with AM should also ensure a high degree of uniformity. This is because such uniformity enables suitable interstitial flow to carry mediators and cells responsible for the degradation of the material. Uniform small pores can thus collect many infiltrates and mediators, which is particularly critical when the microstructure does not allow high flow due to the flow (Corney & Torres-Sanchez, 2008).

5.3.1.3 Environmental factors

The biological environment of the FGBM may also affect their degradation. The pH, temperature, presence of enzymes, and local cellular activities all play important roles in this regard.

- **pH:** The pH of the tissue surrounding the FGBMs with AM may cause or accelerate their degradation. For example, while the degradation of some polymers by hydrolysis reaction might be slow in a normal pH, inflammatory responses may locally create acidic environments that hasten the chemical process.

- **Enzymatic activity:** Cells may produce enzymes that target specific FGBM materials and from the outer environment; this relationship resulted in enzymatic degradation.
- **Cellular activity:** Cellular activities may directly lead to the disintegration of the FGBMs with AM. For example, cells including macrophages and osteoclasts phagocytose particles and secrete enzymes.

In summary, biodegradation is a crucial aspect of any functionally graded biomaterial, as it determines the suitability and efficacy of FGBMs with AM in biomedical applications. However, the ability to control and predict the degradation process makes it possible to use FGBMs with AM that comprise factors such as the material selectivity, porosity, and integrity of the scaffold, as well as various external factors, including biodegradation speed and spatial heterogeneity. Thus, it would be possible to engineer FGBMs with AM that perform complementary roles, allowing them to provide temporary structural support and gradual loading of the regenerating tissue, which may advance the healing process and decrease the need for multiple post-interventional procedures. Overall, the potential to predict biodegradation properties opens new opportunities for utilizing FGBMs with AM in tissue engineering, regenerative medicine, and other biomedical contexts (Murr et al., 2010).

5.3.2 Biodegradation mechanisms

Biodegradation of functionally graded biomaterials involves multiple processes that develop as a result of an interaction in which material is disassembled and assimilated by various biological systems. The mechanisms of materialization of these processes are paramount for engineering generations of FGBMs with AM with decompositional rates atom by atom, which is required for a particular type of biomedical activity. The following three mechanisms of biodegradation are to be discussed in this respect: hydrolytic degradation, enzymatic degradation, and cellular activities.

Table 5.3 summarizes the processes involved in biodegradation, offering a comparative overview for researchers exploring material selection.

5.3.2.1 Hydrolytic degradation

The biomaterials, particularly polymers, degradation mechanism through water presence is frequently through hydrolysis. This degradation type targets the breaking of chemical bonds within the material structure, and water is a catalyst for the scission of this bond. The hydrolysis process through water involves:

- **Polymer backbone cleavage:** Hydrolysis is focused on the ester, amide, or ether groups in the polymer backbone. In hydrolytically degradable polymers such as polylactic acid and polyglycolic acid, ester groups

Table 5.3 Biodegradation mechanisms of FGBMs in AM

Mechanism	*Key features*
Hydrolytic degradation	Breakdown via water catalysis targeting polymer backbones.
Enzymatic degradation	Enzyme-catalyzed disassembly of specific chemical bonds.
Cellular activity	Phagocytosis by immune cells to remove foreign materials.
Polymer degradation	Hydrolysis of ester or ether groups in polymers like PLA and PGA.
Ceramic degradation	Resorption by osteoclasts, especially in calcium phosphates.
Composite behavior	Gradual degradation of composite layers for targeted support.
pH influence	Acidic conditions accelerate hydrolytic degradation.
Porosity	Higher porosity promotes faster integration and degradation.
Environmental factors	Impact of temperature, enzymes, and cellular activity on degradation rate.
Temporal control	Layer-specific degradation to match tissue regeneration.

are hydrolyzed, leading to the degradation of the polymer chain into smaller fragments.

- **pH and temperature:** The rate of hydrolysis is largely affected by other variables such as pH and temperature. Alkaline or acidic conditions increase the rate of hydrolysis while neutral conditions delay it. The rate also increases with a rise in temperature since it makes the molecules more active.
- **Surface erosion:** Hydrolytically degradable polymers degrade initially at the material surface, causing surface erosion. Then, the degradation continues to the inside and causes bulk material erosion as all material undergoes degradation simultaneously.

5.3.3 Enzymatic degradation

Enzymatic degradation refers to the catalyzed breakdown of biomaterials by the enzymes generated by living cells. Enzymes are highly specific in their target and interact only with some chemical bonds or structures within the material. Further, the sub-properties of enzymatic degradation are:

- **Enzymes-substrate interaction:** As mentioned earlier, enzymes such as proteases, lipases, and collagenases break down the specific chemical moieties present within the biomaterials.
- **Cellular production:** Living cells also produce these enzymes, particularly cells of the immune system such as macrophages and neutrophils.
- **Tissue-specific activity:** The tissue environment can also play a role in the degradation activity. In a given tissue, the degradation activity may vary due to the chemical environment such as inflammatory or wound- healing sites.

5.3.4 Cellular activity

Cellular activity is a crucial aspect of the biodegradation process of FGBMs with AM, especially in tissues characterized by high turnover rates or immune responses.

- **Phagocytosis:** Phagocytosis involves the internalization of the foreign particles, which may degrade within the forming phagolysosomes of the phagocytic cell. These include macrophages, which are cells of the immune system specializing in immune debris degradation.
- **Osteoclast activity:** Another potential method of FGBMs with AM degradation is osteoclast activity. Osteoclasts degrade and absorb bone in preparation for remodeling, a process known as resorption. In bone tissue, osteoclast-like cells are synthesized during biomaterial implantation, and debris may be resorbed, including calcium-based ceramics such as hydroxyapatites (Enab, 2014).
- **Inflammatory response:** The inflammatory response is based on the cytophysiologic reaction, and it is involved in the degradation. It enhances cell lysosomal and phagocytic action in the degradation of the periodontium. Besides, this parameter is responsible for demineralization and collagen. The fragmenting matrix is increased by the protease produced by cells. In addition, the inflammatory cells release reactants such as reactive oxygen species or cytokines and release material that affects cell sourcing and enzyme production. The reactants are, for instance, the reactive oxygen species that actually are associative of the oxidative degradation. Cytokines induce cellular activities and affect enzyme production.

To summarize, the biodegradation determinants or processes hydrolytic, enzymatic, and cellular degradation have exerted a primary effect on the behavior of FGBMs with AM and its constituting materials in the biological environment. Through the consideration of these processes and the architectural relationship presented, scientists can design FGBMs with AM with appropriate degradation capacities and adapt the for specific tissue regeneration timelines. Thus, it facilitates the development of biodegradable biomaterials supporting regeneration, restoration, and repair to the host tissue and safe degradation with reduced mitigations or adverse effects.

5.4 APPLICATIONS OF FGBMS WITH AM IN BIOMEDICINE

This allows for a smooth transition to circular economy practices through the use of functionally graded biomaterials (FGBMs), working in conjunction with additive manufacturing (AM). Circular economy principles emphasize re-using, recycling and recovering materials to reduce waste and

Table 5.4 Applications of FGBMs in biomedicine

Application	*Benefits*
Bone tissue engineering	Enhanced mechanical properties; promotes osteogenesis.
Dental implants	Improved osseointegration and aesthetic compatibility.
Skin regeneration	Porous scaffolds for dermal regeneration.
Cartilage repair	Mimics cartilage mechanics for effective chondrogenesis.
Ligament reconstruction	Anisotropic properties ensure tissue alignment and organization.
Vascular tissue engineering	Supports endothelialization for vascular regeneration.
Drug delivery systems	Controlled drug release with minimal side effects.
Patient-specific implants	Custom fabrication for individual anatomical needs.
Circular economy	Recyclable and biodegradable materials for sustainable healthcare.
Smart materials	Responsive behavior for personalized therapeutic applications.

preserve the natural environment. Integrating AM with the latest technology such as biodegradable and recyclable materials, allows industries and businesses to be more sustainable and efficient in terms of resources.

Table 5.4 highlights the specific uses of FGBMs, emphasizing their versatility and relevance in modern healthcare.

5.4.1 Bone tissue engineering

Biodegradable FGBMs reduce the waste produced by medical implants, consistent with a circular economy model. The temporary scaffolds biodegrade after they have served their purpose, negating the need for retrieval surgeries and minimizing long-term medical waste. Additionally, much of the waste produced in AM fabrication is recyclable or reusable, improving lifecycle management of materials in the healthcare field.

Bone tissue engineering is an interdisciplinary field that aims to repair or replace damaged bone tissue by utilizing biocompatible scaffolds that mimic the natural extracellular matrix. Mechanical support is provided by functionally graded biomaterials and cell attachment and proliferation, as well as the direction of new tissue development (Sánchez-Salcedo et al., 2008). These properties of FGBMs with AM make it appropriate for bone tissue engineering:

- Bone tissue engineering- enhanced mechanical properties,
- Osteoconductive properties- adjusted degradation,
- Possibility for patient-specific implants,
- Clinical applications.

Enhanced mechanical properties: To satisfy the mechanical needs of bone tissue, FGBMs with AM can be fabricated with gradient variation in composition and architecture to mimic the mechanical features of native bone tissue. As a result, FGBMs with AM should handle the physiological loads required to perform daily activities.

Osteoconductive properties: FGBMs with AM may also have a bioactive element, such as hydroxyapatite or growth factors. Along with osteoinduction, a bioactive aspect enhances osteoconduction — the ability to bind osteoblastic cells and generate bone.

Controlled degradation: Finally, FGBMs with AM can be designed with controlled degradation; therefore, they gradually lose mass and the mechanical properties of the bone scaffold. This factor is of crucial importance when characterizing and mimicking the healing tissue as the constantly weaker implant does not permit excessive load and transfer the load-bearing to the neo-tissue.

Patient-specific implants: Moreover, FGBMs with AM can be applied for patient-specific implants produced with innovative methods, like additive manufacturing (Zargarian et al., 2016).

Clinical applications: Furthermore, FGBMs with AM have found multiple applications in the clinic as fillings for bone defects or to support the fusion or osseointergration of bone, among others.

Future directions: Current research on the bone tissue engineering of FGBMs with AM is focused on optimizing the scaffolds, fabrication techniques, and increasing biological activity to further observe positive outcomes in clinical settings. Due to material science and regenerative medicine developments, FGBMs with AM are the future in the treatment of bone-related injuries and diseases (Yang et al., 2018).

5.4.2 Dental implants

Dental implants represent one of the pillars of modern dentistry, providing a permanent and effective solution to the lost teeth. Functionally graded biomaterials have recently gained prominence as a viable approach for improving dental implant properties and increasing their longevity (Taheri et al., 2018). Additionally, it seems that FGBMs with AM could address most of the challenges associated with their use, such as absence of optimized osseointegration, biomechanical compatibility, and long-term stability. Other potential benefits associated with the use of FGBMs with AM are as follows:

- **Optimized osseointegration:** FGBM could be engineered depending on the natural gradient of bone tissue, supporting the transition from a mineralized region surrounding a tightly fitting implant to a progressively porous osteoconductive matrix, thus ensuring the correct osseointegration and secure implant anchorage.
- **Enhanced biocompatibility:** FGBMs with AM surface chemistry and topography could create the required cellular responses to enhance adhesion, proliferation, and differentiation or osteogenic

cells, thus enhancing biocompatibility. As a result, the incidence or inflammation-related complications and implant rejection would significantly drop (Lodererova et al., 2009).

- **Custom mechanical properties:** FGBMs with AM provide the ability to modify mechanical properties along the length of the implant to be as close as possible with natural tooth stiffness and resilience, thus limiting the risk of both stress shielding and implant failure due to fractures (Iwasaki, 1997).
- **Reduced peri-implant complications:** All of these changes could contribute for the implants' longer lives and additional comfort for patients. The risk of sub-gingival migration and its accompanying complications would be hence reduced.
- **Aesthetic considerations:** FGBMs with AM could also be modified to resemble natural teeth as closely as possible in terms of color, translucency, and surface texture. It would significantly contribute to the appearance of the final dental restoration and patient satisfaction (Qasim et al., 2020).
- **Advanced surface modification:** FGBMs with AM could also include bioactive coatings, growth factors, or antibacterial compounds to maintain tissue integration and accelerate healing while preventing bacterial colonization.

5.4.3 Soft tissue engineering

Soft tissue engineering is concerned with the restoration and replacement of non-mineralized tissues in the body, including the skin, cartilage, ligaments, and blood vessels, among others. Functionally graded biomaterials prove particularly useful in facilitating such activities and achieving scaffolds with customized properties that are able to imitate the extracellular matrix and promote the targeted tissue regeneration (Wang et al., 2018). The following applications of FGBMs with AM are particularly relevant in soft tissue engineering:

5.4.3.1 Skin regeneration

Long-lasting wounds, including chronic wounds, burns, and severe injuries, often necessitate a skin graft or a tissue-engineered substitute to aid in healing and prevent infection. FGBMs with AM can be bioengineered to imitate the hierarchical organization and mechanical properties of native skin, resulting in cell adhesion, proliferation, and differentiation.

- **Porous scaffolds:** FGBMs with AM with a porosity structure can act as dermal regeneration scaffolds, allowing cells to infiltrate blood vessels and exchange vitamins. Mechanical properties of varying pore

scale and size can integrate tissue growth and encourage the formation of workable layers (Li et al., 2019).
- **Bioactive components:** FGBMs with AM that are supplemented with material that is essential to install the growth factors, cytokines, and extracellular matrix proteins into FGBMs with AM promote cell migration, angiogenesis, and collagen deposition, accelerating wound healing and tissue regeneration.

5.4.3.2 *Cartilage repair*

Treatment of cartilage injuries and degenerative conditions such as osteoarthritis remain challenging due to cartilage's limited regenerative capacity. FGBMs with AM have emerged as promising tools for cartilage repair due to their ability to act as scaffolds similar to native cartilage, thus encouraging differentiation into the chondrogenic lineage.

- **Mechanical Properties:** The mechanical properties can be engineered to replicate those of native cartilage and provide mechanical support to avoid overloading the newly forming tissue.
- **Biomimetic Scaffolds:** The scaffold's stiffness degrades gradually along the scaffold, which aids in cell maturation and extracellular matrix production. biomimetic scaffolds. It can mirror the composition and structure of native temporal barrier cartilage and include bioactive molecules and growth factors to stimulate chondrogenesis and support the development of neohyaline cartilage.

5.4.3.3 *Ligament and tendon reconstruction*

Ligament and tendon injuries often require surgical intervention since they cannot heal adequately to restore function and stability. FGBMs with AM revolutionize ligament and tendon reconstruction by offering scaffolds with a structure that replicates that of native tissues and with appropriate mechanical properties.

- **Anisotropic mechanical properties:** FGBMs with AM are made in such a way that their mechanical properties can be manipulated to be anisotropic. Anisotropy is essential in the orientation and alignment of ligaments and tendons which promotes cell alignment and tissue organization. This fact in biomechanics is key for optimizing graft function (Daxini & Prajapati, 2014).
- **Bioactive biomaterials modification:** Surface modification of FGBMs with AM with biomolecules or growth factors can make the scaffolds bioactive. Bioactive FGBMs with AM increase the adhesion, proliferation, and matrix production of cells, which is essential for the healing process.

5.4.3.4 *Vascular tissue engineering*

Functional blood vessels are required to transport oxygen and nutrients for the optimal functioning of tissues and organs; hence, vascular tissue engineering has emerged as a vital area of investigation. Novel pathways for vascular regeneration include FGBMs with AM, as these scaffolds have been engineered with tailored properties and guidance cues that promote endothelial cell proliferation and vascular network formation.

- **Vascular mimicry:** FGBM can achieve vasculature-like hierarchical structure and mechanical properties that enable for endothelialization on the scaffold and stimulate the formation of functional vascular networks throughout the entire thickness of the scaffold.
- **Biodegradable scaffolds:** FGBM with controlled short- to long- time degradation patterns enable gradual remodeling and natural tissue replacement, thereby ensuring long-lasting vascular network stability and function (Nogata, 1997).

FGBMs with AM have prospective applications in soft tissue (ST) engineering, as they can serve as customizable scaffolds for restoring native-like tissues in the ST compartments. FGBMs with AM possess properties that simulate those of the natural extracellular matrix (EC) and thus enhance cell attachment, growth, and differentiation processes, thus accelerating tissue repair and restoration. Given the ongoing development of ST engineering, FGBMs with AM demonstrate their potential to contribute to solving current clinical issues and enhancing patients' quality of life due to vascular medicine (VM) approaches.

5.4.4 Drug delivery systems

Functionally graded biomaterials have been suggested as ideal candidates for drug delivery vehicles due to the unprecedented control offered by these materials over release kinetics, spatial distribution, and preferential targeting of therapeutics. Through the employment of real or virtually occurring gradients of materials composition, structure, or surface interactions, FGBMs with AM can find applications in virtually all fields of medicine, significantly increasing and further securing the cyberspace of drug delivery.

- **Controlled release kinetics:** FGBMs with AM can be engineered to release drugs in desired kinetics, enabling prolonged action and reduced variability in the drug fluctuations within the body. This feature is critical for drugs that have to maintain steady-state plasma drug levels and those intended for chronic usage.
- **Spatial distribution:** Gradients of material properties enable spatial control of the drug release within the body, thus allowing targeted

drug delivery to specific tissues or organs without off-target effects. This leads to a reduction of undesired drug effects without undermining therapeutic action.

- **Targeted delivery:** FGBMs with AM can be targeted towards delivering drugs only to the site of diseased tissues or cells by addition of targeting ligands or responsive elements. This can not only maximize the therapeutics but also reduce peripheral side effects and systemic toxicity.
- **Multifunctional platforms:** FGBMs can also act as a multifunctional platform to deliver more than one drug or therapeutic agent. A material can comprise regions with different drug loading capacities in a single system, multiple agents in the same material or a region that uses therapeutic effects synergistically.
- **Biodegradability and biocompatibility:** Being biodegradable, FGBMs with AM degrade over time into biocompatible byproducts that avoid the surgical removal and prevent foreign body responses. Since the biocompatibility is important for long-term implantation and the sustained drug delivery, many of the FGBM applications verge on the development of FGBMs for drug delivery.
- **Versatility and customization:** FGBMs with AM are versatile and customizable, which means that they allow developing the drug delivery systems tailored to the specific therapeutic application. The drug release can be controlled by the material properties and by the configuration of the drug loading, and resistance profiles can be adjusted to satisfy the diverse clinical needs.
- **Emerging technologies:** The possibilities of the emerging technologies, such as nanotechnology and biomaterials engineering, created the opportunity for developing the novel FGBM-based drug delivery platforms. These encompass nanoparticles, microparticles, hydrogels, and implantable devices, among others, and show the great promise for targeted, personalized, and minimally invasive drug delivery approaches.

Functionally graded biomaterials have been recognized as a viable foundation for drug delivery systems in biomedicine. They provide unprecedented opportunities to control drug release kinetics and targeting through systematic creation of gradients in material properties. As a drug delivery platform, FGBMs with AM have the potential to improve the efficacy, safety, and scope of use in drug delivery. Nevertheless, much more research and development of novel approaches is necessary to keep this driving force ahead in the future.

5.4.5 Circular economy applications in broader sectors

AM-based FGBMs are finding applications beyond biomedicine, with the fields of architecture and agriculture being just some of the areas showing interest. AM techniques can further address the application of recycled polymers and natural fibers towards producing sustainable construction materials, agricultural equipment, etc. This allows both techniques to minimize raw material usage and guarantees that waste from one procedure can become feedstock for another—encapsulating a closed-loop manufacturing framework.

5.5 FUTURE DIRECTIONS AND RESEARCH OPPORTUNITIES

Based on the above challenges and opportunities, we believe that the field of FGBMs with AM has just scratched the surface, and numerous avenues of exciting research exist within it. Thus, the abovementioned potential includes:

- **Advanced fabrication techniques:** AM and electrospinning provide new opportunities to prepare complex FGBMs with AM with controlled composition, structure and properties (Anné et al., 2006).
- **Bioactive functionalization:** Including incorporation of bioactive molecules, growth factors, and signaling cues into FGBMs with AM to aid tissue regeneration, cell differentiation, and modulation of the immune responses.
- **Smart and responsive materials:** responsive FGBMs with AM composed of stimuli-responsive polymers and drug-releasing coatings, etc. in order to enhance drug release, change material properties on command, and realize true personalized medicine.
- **In vivo evaluation and clinical translation:** Preclinical and clinical studies are integral part of the development process in order to investigate FGBM safety, efficacy, and long-term performance through rigorous testing in relevant animal models and humans.
- **Multimodal imaging and monitoring technologies:** For example, integration of imaging techniques and biological sensors to monitor the tissue regeneration, self-regulated drug delivery, and patient status in real time.

Additive manufacturing energy efficiency is crucial for environmental sustainability. Utilizing AM- related methods like the optimized printing parameters and optimization of topology can enable product production in low energy consumption mode. AM allows for lightweight designs with reduced materials consumption which can decrease the energy spent on transporting and assembling parts across many application areas, most notably in the aerospace and automotive industries.

Table 5.5 outlines ongoing and potential research areas, guiding researchers toward innovative applications and advancements in FGBMs.

Table 5.5 Emerging research directions for FGBMs in AM

Research area	*Focus*
Advanced fabrication techniques	Development of AM methods like electrospinning and multi-material printing.
Bioactive functionalization	Incorporation of growth factors, signaling molecules, and coatings.
Smart materials	Integration of responsive polymers for adaptive biomedical applications.
Multimodal imaging integration	Real-time monitoring of tissue regeneration and implant behavior.
Environmental sustainability	Use of renewable and biodegradable materials in AM processes.
Drug delivery innovations	FGBMs as multifunctional platforms for targeted drug release.
Personalized medicine	Customized implants and devices for individual patient anatomy.
Long-term biocompatibility studies	Evaluation of chronic effects of FGBMs in living systems.
Enhanced scaffold designs	Optimization of porosity and mechanical properties for tissue engineering.
Regulatory and ethical challenges	Addressing approval and safety concerns in FGBM clinical applications.

5.5.1 Potential impact on medical treatments and patient outcomes

The potential applications of functionally graded biomaterials in medicine and regenerative therapeutics far exceed the scope of current uses. In particular, widespread adoption of such biomaterials has several advantages, which are summarized as follows:

- **Enhanced regenerative therapies:** FGBMs with AM make it achievable to engineer complex tissues and organs by allowing for a custom approach, which promises the high functionality and integration of developed structures.
- **Minimally invasive interventions:** Novel drug delivery systems based on FGBMs with AM make it possible to design interventions that minimize activity of required areas. It further enables physicians to avoid some surgeries or the administration of specific drugs through the systemic circulation.
- **Personalized medicine:** Due to the fact that FGBMs with AM make it feasible to approach the treatment process from a personalized perspective, costs related to incorrect drug and treatment choices will be reduced; the cost of care will decrease, and patient outcomes and satisfaction will improve.

- **Extended device lifespan:** If biodegradable FGBMs with AM are utilized, implants will predictably not require removal. Instead, they serve as temporary supports, which degrade as tissues regenerate.

By minimizing the carbon impact of manufacturing processes, AM is consistent with global sustainability goals due to its energy efficiency. This is significant in biomedical applications since energy-heavy processes can be supplanted with AM-based procedures that permit modification and tweaking of inserts or gadgets with a minimal ecological impact.

5.6 CONCLUSION

Exploring the application of functionally graded biomaterials (FGBMs) in additive manufacturing (AM) illuminates the true potential of their potential to revolutionize biomedicine for tissue engineering, regenerative medicine, and drug delivery. Owing to their distinct material properties and tunable function, FGBMs represent a novel strategy for medical practitioners, with a goal of enhancing patient outcomes. Diverse surface chemistry, mechanical properties, and compositional gradients ensure the biocompatibility of FGBMs and efficient incorporation in biological systems. Moreover, the hydrolytic and enzymatic degradation processes of FGBMs result in the gradual replacement of tissues via regeneration. Their versatility to meet various clinical demands is demonstrated by their effective use in fields such as bone and soft tissue engineering, dental implants, and drug delivery. Bold improvements in AM fabrication strategies, bioactive functionalization and the emergence of smart materials will be the key players for a new era in the FGBMs evolution, opening new avenues for precision medicine and regenerative therapeutics. They work to integrate these materials with biomedical imaging and personalized drug therapies, which could revolutionize medical practices and the way patient care is performed. Overall, the ongoing research and development of FGBMs in AM opens up a promising future for biomedicine, where genuine patient-centered healthcare can be achieved, with the opportunity to add an empathetic touch to patient care in line with the ethos of contemporary healthcare specifically, patient-centered care.

REFERENCES

Achrai, B., & Wagner, H. D. (2017). The turtle carapace as an optimized multi-scale biological composite armor – A review. *Journal of the Mechanical Behavior of Biomedical Materials*, *73*, 50–67. https://doi.org/10.1016/j.jmbbm.2017.02.027

Afzal, M. A. F., Kesarwani, P., Reddy, K. M., Kalmodia, S., Basu, B., & Balani, K. (2012). Functionally graded hydroxyapatite-alumina-zirconia biocomposite: Synergy of toughness and biocompatibility. *Materials Science and Engineering C*, *32*(5), 1164–1173. https://doi.org/10.1016/j.msec.2012.03.003

Anné, G., Vanmeensel, K., Vleugels, J., & Van Der Biest, O. (2006). Electrophoretic deposition as a novel near net shaping technique for functionally graded biomaterials. *Key Engineering Materials, 314*, 213–218. https://doi.org/10.4028/0-87849-998-9.213

Bahraminasab, M., Arab, S., & Doostmohammadi, N. (2022). Cytotoxicity and ion release of functionally graded Al_2O_3-Ti orthopedic biomaterial. *Journal of Biomimetics, Biomaterials and Biomedical Engineering, 54*, 103–118. https://doi.org/10.4028/www.scientific.net/JBBBE.54.103

Bambole, V., & Yakhmi, J. V. (2016). Tissue engineering: use of electrospinning technique for recreating physiological functions*Nanobiomaterials in Soft Tissue Engineering* (pp. 387–455). Elsevier. https://doi.org/10.1016/B978-0-323-42865-1.00014-3

Bogdanski, D., Koller, M., Muller, D., Muhr, G., Bram, M., Buchkremer, H. P., Stover, D., Choi, J., & Epple, M. (2002). Easy assessment of the biocompatibility of Ni-Ti alloys by in vitro cell culture experiments on a functionally graded Ni-NiTi-Ti material. *Biomaterials, 23*(23), 4549–4555. https://doi.org/10.1016/S0142-9612(02)00200-4

Corney, J. R., & Torres-Sanchez, C. (2008). Towards functionally graded cellular microstructures. *Proceedings of the ASME Conference on Smart Materials, Adaptive Structures and Intelligent Systems, SMASIS2008, 1*, 45–53. https://doi.org/10.1115/SMASIS2008-414

Davoodi, E., Montazerian, H., Mirhakimi, A. S., Zhianmanesh, M., Ibhadode, O., Shahabad, S. I., Esmaeilizadeh, R., Sarikhani, E., Toorandaz, S., Sarabi, S. A., Nasiri, R., Zhu, Y., Kadkhodapour, J., Li, B., Khademhosseini, A., & Toyserkani, E. (2022). Additively manufactured metallic biomaterials. *Bioactive Materials, 15*, 214–249. https://doi.org/https://doi.org/10.1016/j.bioactmat.2021.12.027

Daxini, S. D., & Prajapati, J. M. (2014). A review on recent contribution of meshfree methods to structure and fracture mechanics applications. *The Scientific World Journal, 2014*. https://doi.org/10.1155/2014/247172

Dorozhkin, S. V. (2010). *Calcium Orthophosphate-Based Biocomposites and Hybrid Biomaterials*. Hauppauge, NY: Nova Science Publishers, Inc.

Emir, E., & Bahçe, E. (2023). Investigation of the mechanical properties triple periodic minimal surfaces lattice structures with functional graded of porosity. *Journal of Cellular Plastics, 59*(4), 251–268. https://doi.org/10.1177/0021955X231175179

Enab, T. A. (2012). A comparative study of the performance of metallic and FGM tibia tray components in total knee replacement joints. *Computational Materials Science, 53*(1), 94–100. https://doi.org/10.1016/j.commatsci.2011.09.032

Enab, T. A. (2014). Performance improvement of total knee replacement joint through bidirectional functionally graded material. *International Journal of Mechanical and Mechatronics Engineering, 14*(2), 104–113.

Geoghegan, N., O'Loughlin, M., Delaney, C., Rochfort, K. D., Kennedy, M., Kolagatla, S., Podhorska, L., Rodriguez, B. J., Florea, L., & Kelleher, S. M. (2023). Controlled degradation of polycaprolactone-based micropillar arrays. *Biomaterials Science, 11*(9), 3077. https://doi.org/10.1039/D3BM00165B

Gibson, A. L., Wilson, L. H., Wilkerson, W. R., Feinberg, A. W., Seegert, C. A., Baney, R. H., & Brennan, A. B. (2002). Characterization of chemically and topographically modified siloxane elastomer for controlled cell growth. *Materials Research Society Symposium - Proceedings, 711*, 169–174.

Goto, T. (2015). Functionally graded materials biomaterials. *Funtai Oyobi Fummatsu Yakin/Journal of the Japan Society of Powder and Powder Metallurgy*, *62*(8), 390. https://doi.org/10.2497/jjspm.62.390

Goto, T. (2018). Functionally graded materials · Biomaterials. *Funtai Oyobi Fummatsu Yakin/Journal of the Japan Society of Powder and Powder Metallurgy*, *65*(2), 79. https://doi.org/10.2497/jjspm.65.79

Gupta, D., Chaudhary, A. K., Singh, V. K., Verma, D., Goh, K. L., & Sharma, M. (2023). Thermo-mechanical analysis of bhimal fiber (*Grewia optiva*)-$CaCO_3$/flyash/TiO_2 reinforced epoxy bio-composites. *Industrial Crops and Products*, *204*, 117341. https://doi.org/https://doi.org/10.1016/j.indcrop.2023.117341

Gupta, D., Dogra, V., Verma, D., Chaudhary, A. K., & Tewari, M. (2024). Chapter 4- PCL-based composites and their utilizations in the medical sector. In D. Verma, M. Okhawilai, K. L. Goh, S. Ramakrishna, P. Pasbakhsh, & M. Sharma (Eds.), *Woodhead Publishing Series in Biomaterials* (pp. 63–83). Woodhead Publishing. https://doi.org/10.1016/B978-0-443-18915-9.00012-4

Hedia, H. S., El-Midany, T. T., Shabara, M. A. N., & Fouda, N. (2006). Design optimization of cementless metal-backed cup prostheses using the concept of functionally graded material. *Biomedical Materials*, *1*(3), 127–133. https://doi.org/10.1088/1748-6041/1/3/006

Hedia, H. S., Shabara, M. A. N., El-Midany, T. T., & Fouda, N. (2006). Improved design of cementless hip stems using two-dimensional functionally graded materials. *Journal of Biomedical Materials Research - Part B Applied Biomaterials*, *79*(1), 42–49. https://doi.org/10.1002/jbm.b.30509

Iwasaki, K. (1997). Production of a functionally graded artificial tooth root by unique sequence of processes. *Materials Research Innovations*, *1*(3), 180–187. https://doi.org/10.1007/s100190050038

Kabiraj, A., Halder, U., Panja, A. S., Chitikineni, A., Varshney, R. K., & Bandopadhyay, R. (2023). Detailed genomic and biochemical characterization and plant growth promoting properties of an arsenic-tolerant isolate of Bacillus pacificus from contaminated groundwater of West Bengal, India. *Biocatalysis and Agricultural Biotechnology*, *52*, 102825. https://doi.org/10.1016/j.bcab.2023.102825

Leong, K. F., Chua, C. K., Sudarmadji, N., & Yeong, W. Y. (2008). Engineering functionally graded tissue engineering scaffolds. *Journal of the Mechanical Behavior of Biomedical Materials*, *1*(2), 140–152. https://doi.org/10.1016/j.jmbbm.2007.11.002

Li, S., Jiang, M., Zhang, Y., Xie, X., Li, W., Ming, P. Y., Jiang, X., Yang, B., He, Y., Chen, J., & Tao, G. (2023). Multi-functional carboxymethyl chitosan/sericin protein/halloysite composite sponge with efficient antibacterial and hemostatic properties for accelerating wound healing. *International Journal of Biological Macromolecules*, *234*, 123357. https://doi.org/10.1016/J.IJBIOMAC.2023.123357

Li, Y., Jahr, H., Pavanram, P., Bobbert, F. S. L., Puggi, U., Zhang, X.-Y., Pouran, B., Leeflang, M. A., Weinans, H., Zhou, J., & Zadpoor, A. A. (2019). Additively manufactured functionally graded biodegradable porous iron. *Acta Biomaterialia*, *96*, 646–661. https://doi.org/10.1016/j.actbio.2019.07.013

Lin, D., Li, Q., Li, W., & Swain, M. (2010). Bone remodeling induced by dental implants of functionally graded materials. *Journal of Biomedical Materials Research - Part B Applied Biomaterials*, *92*(2), 430–438. https://doi.org/10.1002/jbm.b.31531

Lodererova, M., Rybnicek, J., Steidl, J., Richter, J., Boivie, K., Karlsen, R., & Åsebø, O. (2009). Biocompatibility of metal sintered materials in dependence on multi-material graded structure. *IFMBE Proceedings*, *23*, 1204–1207. https://doi.org/10.1007/978-3-540-92841-6_295

Lvov, Y., Wang, W., Zhang, L., & Fakhrullin, R. (2016). Halloysite clay nanotubes for loading and sustained release of functional compounds. *Advanced Materials*, *28*(6), 1227–1250. https://doi.org/10.1002/ADMA.201502341

Ma, J., Goble, K., Smietana, M., Kostrominova, T., Larkin, L., & Arruda, E. M. (2009). Morphological and functional characteristics of three-dimensional engineered bone-ligament-bone constructs following implantation. *Journal of Biomechanical Engineering*, *131*(10). https://doi.org/10.1115/1.4000151

Madeira, S., Souza, J. C. M., Fredel, M. C., Henriques, B., Silva, F. S., & Zhang, Y. (2018). Functionally graded nanostructured biomaterials (FGNB). In *Nanostructured Biomaterials for Cranio-Maxillofacial and Oral Applications* (pp. 159–180). Elsevier Inc. https://doi.org/10.1016/B978-0-12-814621-7.00009-3

Mohammadi, P., Gandier, J.-A., Wagermaier, W., Miserez, A., & Penttilä, M. (2021). Bioinspired functionally graded composite assembled using cellulose nanocrystals and genetically engineered proteins with controlled biomineralization. *Advanced Materials*, *33*(42). https://doi.org/10.1002/adma.202102658

Murr, L. E., Gaytan, S. M., Medina, F., Lopez, H., Martinez, E., MacHado, B. I., Hernandez, D. H., Martinez, L., Lopez, M. I., Wicker, R. B., & Bracke, J. (2010). Next-generation biomedical implants using additive manufacturing of complex cellular and functional mesh arrays. *Philosophical Transactions of the Royal Society A: Mathematical, Physical and Engineering Sciences*, *368*(1917), 1999–2032. https://doi.org/10.1098/rsta.2010.0010

Naahidi, S., Jafari, M., Edalat, F., Raymond, K., Khademhosseini, A., & Chen, P. (2013). Biocompatibility of engineered nanoparticles for drug delivery. *Journal of Controlled Release : Official Journal of the Controlled Release Society*, *166*(2), 182–194. https://doi.org/10.1016/j.jconrel.2012.12.013

Nogata, F. (1997). Learning about design concepts from natural functionally graded materials. *ASME International Mechanical Engineering Congress and Exposition, Proceedings (IMECE)*, *1997-N*, 11–18. https://doi.org/10.1115/IMECE1997-0646

Prakash, C., & Singh, S. (2020). On the characterization of functionally graded biomaterial primed through a novel plaster mold casting process. *Materials Science and Engineering C*, *110*. https://doi.org/10.1016/j.msec.2020.110654

Qasim, S. S. B., Zafar, M. S., Niazi, F. H., Alshahwan, M., Omar, H., & Daood, U. (2020). Functionally graded biomimetic biomaterials in dentistry: An evidence-based update. *Journal of Biomaterials Science, Polymer Edition*, *31*(9), 1144–1162. https://doi.org/10.1080/09205063.2020.1744289

Sánchez-Salcedo, S., Werner, J., & Vallet-Regí, M. (2008). Hierarchical pore structure of calcium phosphate scaffolds by a combination of gel-casting and multiple tape-casting methods. *Acta Biomaterialia*, *4*(4), 913–922. https://doi.org/10.1016/j.actbio.2008.02.005

Scaffaro, R., Lopresti, F., Botta, L., Rigogliuso, S., & Ghersi, G. (2016). Preparation of three-layered porous PLA/PEG scaffold: Relationship between morphology, mechanical behavior and cell permeability. *Journal of the Mechanical Behavior of Biomedical Materials*, *54*, 8–20. https://doi.org/10.1016/j.jmbbm.2015.08.033

Shah, A. T., Zahid, S., Ikram, F., Maqbool, M., Chaudhry, A. A., Rahim, M. I., Schmidt, F., Goerke, O., Khan, A. S., & Rehman, I. U. (2019). Tri-layered functionally graded membrane for potential application in periodontal regeneration. *Materials Science and Engineering C, 103*. https://doi.org/10.1016/j.msec.2019.109812

Taheri, R. A., Jarrahi, A., Farnoosh, G., & Karimi, A. (2018). A comparative finite element simulation of stress in dental implant–bone interface using isotropic and orthotropic material models in three mastication cycles. *Journal of the Brazilian Society of Mechanical Sciences and Engineering, 40*(10). https://doi.org/10.1007/s40430-018-1409-9

Tirta Nindhia, T. G., Koyoshi, Y., Kaneko, A., Sawada, H., Ohta, M., Hirai, S., & Uo, M. (2008). Hydroxyapatite-silk functionally graded material by pulse electric current sintering. *Trends in Biomaterials and Artificial Organs, 22*(1), 25–29.

Veloso, F., Miranda, D., Morais, P., Torres, H. R., Oliveira, B., Correia-Pinto, J., Pinho, A. C. M., & Vilaça, J. L. (2022). Study of the compression behavior of functionally graded lattice for customized cranial remodeling orthosis. *Journal of the Mechanical Behavior of Biomedical Materials, 130*. https://doi.org/10.1016/j.jmbbm.2022.105191

Wang, C., Zhang, X., Wang, D. M., Yung, P. S. H., Tuan, R. S., & Ker, D. F. E. (2024). Optimized design of an enthesis-mimicking suture anchor-tendon hybrid graft for mechanically robust bone-tendon repair. *Acta Biomaterialia, 176*, 277–292. https://doi.org/10.1016/j.actbio.2024.01.011

Wang, H., Su, K., Su, L., Liang, P., Ji, P., & Wang, C. (2018). The effect of 3D-printed Ti_6Al_4V scaffolds with various macropore structures on osteointegration and osteogenesis: A biomechanical evaluation. *Journal of the Mechanical Behavior of Biomedical Materials, 88*, 488–496. https://doi.org/10.1016/j.jmbbm.2018.08.049

Watanabe, Y., Iwamoto, T., Teramoto, A., Abe, K., & Ohkoshi, Y. (2004). Biodegradability of poly (L-lactic acid) functionally graded materials with hardness gradient. *Materials Transactions, 45*(4), 1005–1009. https://doi.org/10.2320/matertrans.45.1005

Werner, J., Linner-Krčmar, B., Friess, W., & Greil, P. (2002). Mechanical properties and in vitro cell compatibility of hydroxyapatite ceramics with graded pore structure. *Biomaterials, 23*(21), 4285–4294. https://doi.org/10.1016/S0142-9612(02)00191-6

Xiong, X., Qu, S.-X., & Liu, Y.-M. (2013). Synthesis and characterization of Dopamine graft compound *N*-methacryloyl 3,4-dihydroxyl-phenylamine. *Journal of Physics: Conference Series, 419*(1). https://doi.org/10.1088/1742-6596/419/1/012047

Xu, J., Zhao, X., Munroe, P., & Xie, Z. (2014). Synergistic toughening of hard, nacre-mimetic $MoSi_2$ coatings by self-assembled hierarchical structure. *Scientific Reports, 4*. https://doi.org/10.1038/srep04239

Yang, C., Wu, H., Chen, S., & Kang, G. (2018). Three-dimensional bioglass-collagen-phosphatidylserine scaffolds designed with functionally graded structure and mechanical features. *Biomedizinische Technik, 63*(3), 255–259. https://doi.org/10.1515/bmt-2017-0185

Zargarian, A., Esfahanian, M., Kadkhodapour, J., & Ziaei-Rad, S. (2016). Numerical simulation of the fatigue behavior of additive manufactured titanium porous lattice structures. *Materials Science and Engineering C*, *60*, 339–347. https://doi.org/10.1016/j.msec.2015.11.054

Zhu, D., Zhang, C., & Liu, P. (2018). Study on the design of natural and biomimetic flexible biological structures. *Fuhe Cailiao Xuebao/Acta Materiae Compositae Sinica*, *35*(6), 1636–1645. https://doi.org/10.13801/j.cnki.fhclxb.20170920.001

Chapter 6

Applications and future directions of additive manufacturing

Upender Punia and Ramesh Kumar Garg

ABBREVIATIONS

AM	additive manufacturing
FDM	fused deposition modelling
SLS	selective laser sintering
SLM	selective laser melting
AI	artificial intelligence
ML	machine learning
3D	three dimensional
FFF	fused filament fabrication
PBF	powder bed fusion

6.1 INTRODUCTION

3D printing or additive manufacturing (AM) is an innovative production technique that creates parts layer-by-layer from a virtual model (Jiménez et al., 2019). AM is different from other approaches of manufacturing that involve material cutting from a solid mass, as AM only deposits materials in a layer where necessary, thus reducing wastage of material and encouraging the fabrication of complex geometries (Bikas et al., 2016; Jadhav & Jadhav, 2022). This technology is therefore a revolution in the approach or thinking where a product idea is conceived, designed, and prototyped (Tan et al., 2023).

AM has been applied in various sectors since it offers high flexibility and effectiveness (Mehrpouya et al., 2019). In aerospace, it enables the formation of complex thin-walled structures, turbine blades and brackets among others that help to improve efficiency in fuel consumption and other performance-based attributes (Najmon et al., 2019). In the automotive industry, AM is used for prototyping tools, manufacturing of small to medium or part-specific items, including battery boxes for electric vehicles (Mahamood et al., 2021). In the construction industry, AM is used to manufacture parts of structures and even whole objects with the use of lesser manpower than traditional approaches, and with less expenses on materials

 DOI: 10.1201/9781003607052-6

(Delgado Camacho et al., 2018). In energy, AM helps produce power-effective turbine and heat exchange systems; in Education, AM encourages quick idea creation by allowing designing, with a practical model, of new innovative concepts (Motyl & Filippi, 2021; Tofail et al., 2018).

It has been observed that the success of AM relies on the fact that it is capable of operating across a wide range of heterogeneous materials specific to applications (Bhatia & Sehgal, 2023). Thermoplastics are predominant in AM for parts application based on prototypes and consumer products because of their manufacturability and low-cost (Picard et al., 2020). Titanium, aluminum, and stainless steel are used in aerospace, automotive, and power systems industries owing to properties like high strength-to-weight ratios, lightweight, and high endurance (Moeinfar et al., 2022). The costly but high-performing materials such as composites and ceramics are being adopted in AM to realize characteristics such as high thermal stability and low-density strength (Lakhdar et al., 2021). In healthcare, biocompatible materials which include polylactic acid (PLA) and hydroxyapatite are used in the production of implants and for the manufacture of tissue engineering scaffolds (A. B. Singh, 2024). In addition, advanced studies on the use of multi-material systems are allowing the generation of parts with specific properties such as flexible and stiff areas within the same piece (Bandyopadhyay & Heer, 2018). The adoption of AM into these applications has been disruptive. It has reduced product development cycle, cut material costs, and brought mass customization into the picture. Due to advanced manufacturing methods, the geometry of parts has increased the range of design and AM parts, in turn has led to innovations that were not possible with traditional manufacturing processes. Moreover, the localized production capability of AM fits well in manufacturing-on-demand, since there is little dependency on extensive supply chains and ending up with numerous inventories and intermediate stocks (Kunovjanek et al., 2022; Pereira et al., 2019).

Further development of AM seems assured in the same way that it constitutes an integral part of present and future manufacturing as its opportunities and material capabilities are extended. The next parts further discuss the present uses and future potential of AM together with showcasing various successful applications as well as the potential of this technology to revolutionise various industries.

6.1.1 Recent study in field of resin 3D printing

The development of 3D Printing technology has resulted in important breakthroughs in various fields, including the medical, chemical, engineering, environmental, genetics, dentistry, etc. Traditional methods are progressively being superseded by digital fabrication and design procedures in the field of medical applications. Hollow, tiny, complicated, and sophisticated sculptures may be made with this method with astounding precision and accuracy while still being reasonably priced (Javaid & Haleem, 2019; Shaikh et al., 2021).

The number of publications for the last 15 years was searched in the database of PubMed and Scopus by using the keywords [3D Printing]. The results are plotted in form of bar graph in Figure 6.1. The results indicated that the number of publications in this area is increasing abruptly every year. The research in resin 3D Printing is a subject of exploration across various fields, as illustrated in the horizontal bar graphs depicted in Figure 6.2. With ongoing advancements, recent focus has shifted towards the search of novel materials suitable for 3D Printing, as well as exploring its engineering applications.

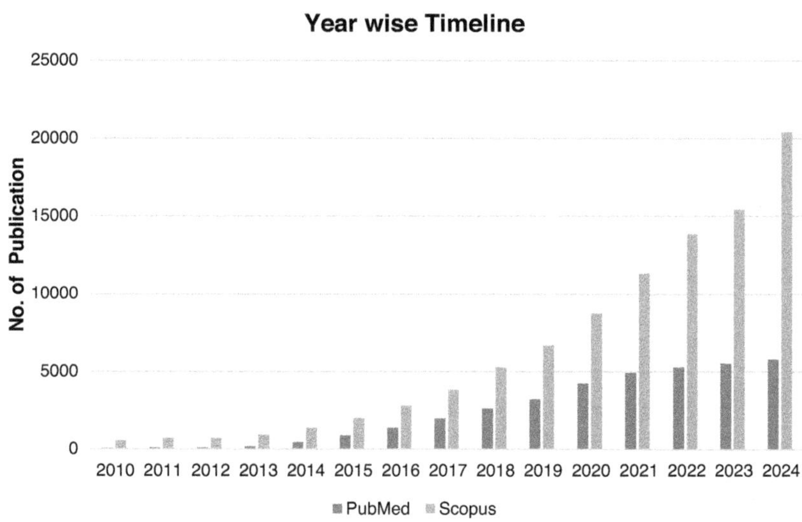

Figure 6.1 Year-wise timeline results of PubMed and Scopus with keywords (3D Printing).

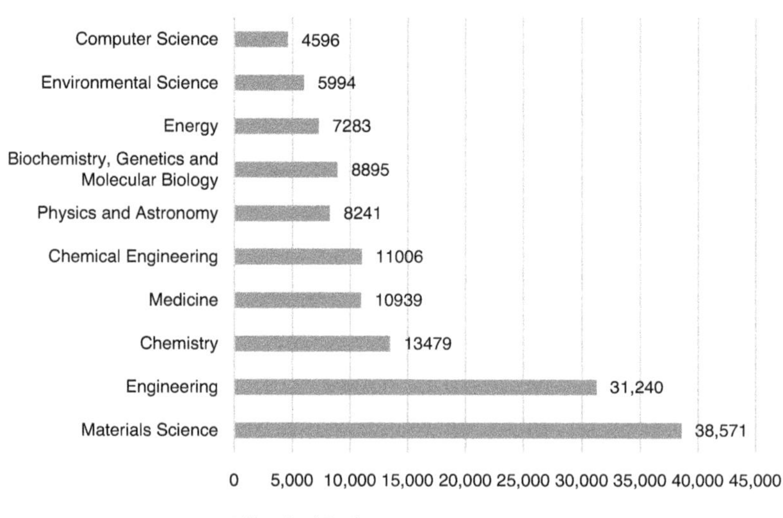

Figure 6.2 Research on 3D Printing in various fields in last 15 years.

6.2 CLASSIFICATION OF AM

6.2.1 Material extrusion

Material extrusion is a common and affordable additive manufacturing (AM) technique. The thermoplastic material is used in this technique (Altıparmak et al., 2022). The material is heated, and when it becomes viscous or in a flowable state, it is extruded on the build plate through the nozzle in a desired pattern. On the release of heat, the material further becomes a solid where the layers stick to the preceding layer. This technique is most closely linked with Fused Deposition Modelling, which is the most commonly used technology for rapid prototyping, educational models, and regular end-use parts (Oleff et al., 2021). In material extrusion, materials to be processed include plastics familiar to most people, like PLA, ABS, PETG, etc., and the cutting-edge materials like carbon fiber-reinforced polymers. The process is suited for producing cheap functional products and lightweight parts and that makes it suitable for small companies and individuals who use 3D Printing for recreation or as a start-up venture (Shanmugam et al., 2021) (Figure 6.3).

6.2.2 Vat photopolymerization

Vat photopolymerization is an accuracy-dependent AM technique that employs liquid photopolymer resin, which is selectively solidified using a

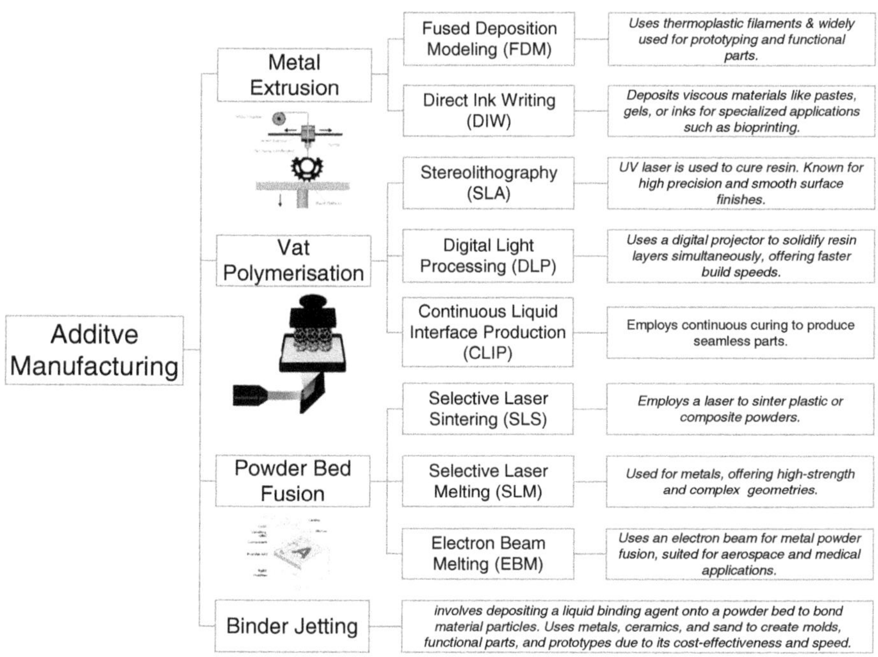

Figure 6.3 Classification of AM.

light source, for instance, a laser or a projector (Davoudinejad, 2021). The main types of this process are stereolithography (SLA), digital light processing (DLP), and continuous liquid interface production (CLIP). These technologies allow for the fabrication of accurate parts and fine surface textures which are ideal for applications in industries such as medical, dental, and in the production of specialized designs such as jewelry (Pazhamannil & Govindan, 2020). The liquid resins may be adjusted in order to obtain specific characteristics, for example, increased stiffness, elasticity, or light permeability, thus extending the potential uses of this method. This has ensured that vat photopolymerization is frequently used in industries that need accurate functional parts and prototypes (Al Rashid et al., 2021).

6.2.3 Powder bed fusion

AM technologies that build up a component from a powder bed by using a laser, an electron beam, or another heat source are referred to as powder bed fusion (PBF) (Scime & Beuth, 2018). Some of the noteworthy techniques that are discussed here are selective laser sintering (SLS) for polymers and composites, selective laser sintering (SLS), direct metal laser sintering (DMLS), and electron beam melting (EBM) for metals. PBF particularly performs well in various geometries that are strong and rigid; it plays an important role in aerospace, healthcare, and automotive e industries. It includes materials like nylon and polyamide to metallic elements such as titanium and aluminum, among others. Due to its high precision, as well as a range of applicable materials, PBF can be used for manufacturing lightweight aerospace components, different medical implants, and function prototypes for sophisticated engineering tasks (Gui et al., 2022; R. Singh et al., 2019).

6.3 APPLICATION OF AM

Traditionally known as rapid prototyping, additive manufacturing (AM) has transformed several industries by allowing for intricate structures, minimizing material wastage, and prototyping (Mpofu et al., 2014). Owing to its multiplicity and low-cost nature, the technology has opened the door for developmental uses in various industries; hence, a few examples are outlined below.

6.3.1 Aerospace industry

AM has dramatically affected the aerospace industry through the creation of complex yet lightweight and custom-designed and developed part which is at once both efficacious and economical (Martinez et al., 2022). Different

AM techniques are applied in aerospace applications and depend on the material type, shape, and mechanical properties needed for that part. Some of the aerospace parts produced through AM are highlighted in this section below. Turbine blades and fuel nozzles are integral components of aircraft engines; they must possess high specific strength, high temperature resistance, and low density. Turbine blades are very effectively produced through direct metal laser sintering (DMLS) using metal powder of Inconel and titanium alloys (Kalender et al., 2019).

An advanced form of rapid prototyping system is gradually adopted in manufacturing the structural parts of the airframe, including wings, ribs, spars, and fuselage parts. These parts include those made from lightweight materials such as aluminum or any titanium alloys having printing techniques such as SLS or SLM. These parts benefit from the capability of building accurate geometries and thinning the material to avoid losing strength. Thus, the resultant parts are not only stronger but lighter than the classical castings or machined spares and the overhead weight of aircraft can be minimized. In aerospace manufacturing, part-specific tools and fixtures are normally required throughout the manufacturing life cycle of product assembly, maintenance, and testing (Joshi & Sheikh, 2015; Wawryniuk et al., 2024). They can be often made using FDM or SLA. Specific types of plastics, such as ABS, nylon, and carbon fiber composites are utilized because the material provides adequate stiffness and rigidity for manipulation during manufacturing. Conventional tooling processes are no longer essential. Each of these parts also benefits from the use of additive manufacturing advantages: lighter, better performance, more versatility, and internal geometry that could not be done otherwise (Angrish, 2014). The type and kind of AM technique and the particular material selected for each part depend on the specifications such as mechanical properties, thermal stability, weight of the part, and performance characteristic of the component in the aircraft system (Mouzakis, 2018) (Figure 6.4).

6.3.2 Healthcare

There is an evident and ever-growing impact from additive manufacturing (AM) within the medical industry for the production of specific tools, especially devices, and implants tailored specifically for a particular patient, as well as the evolution of the surgical instruments and prostheses (Javaid et al., 2022). Different AM techniques are used as per the requirement, and all of them ensure better accuracy, less cycle time, and better results. This section presents a number of components of medical equipments that have been fabricated by AM (Dodziuk, 2016). Orthopedic implants remain one of the most practical realizations of AM, with a high level of individualization of products (Murr, 2020). Selective laser melting (SLM) is particularly employed to build implants through DMLS out

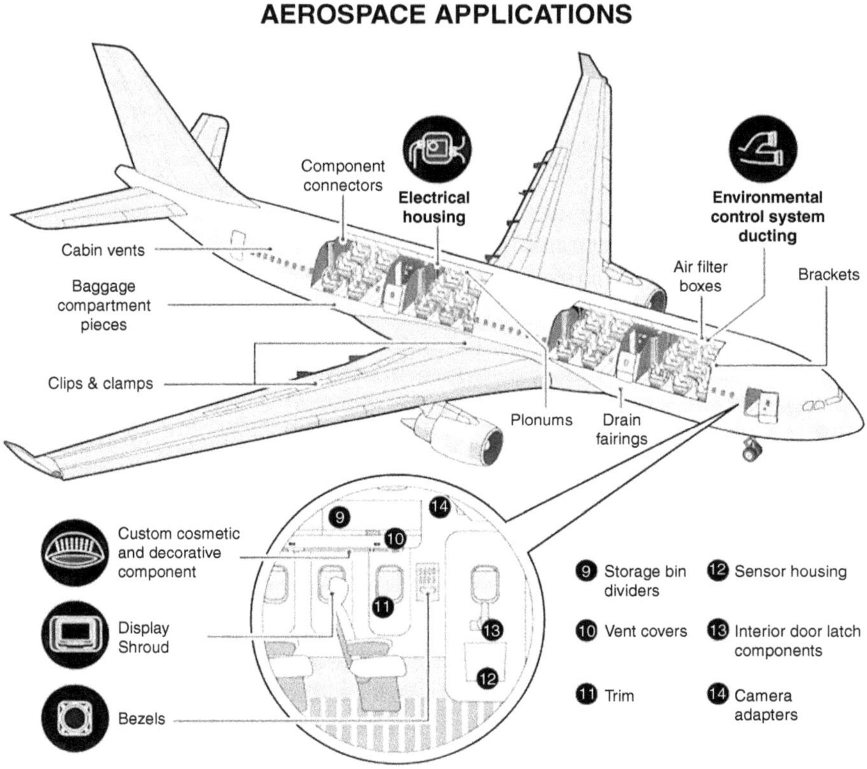

Figure 6.4 Application of AM in aerospace industries.

of biocompatible materials ranging from titanium and its alloys, which are ideal for bone replacement. These implants are usually shaped to fit the patient's body, which eliminates most of the complications associated with them. AM in the creation of implants also enables the production of highly complex structures of a lattice that encourage the implant to bond with the bone tissue (Meng et al., 2023). Artificial limbs, including hands, arms, and legs are produced using Fused Deposition Modelling (FDM) or Stereolithography (SLA) (Rojek et al., 2020). Thus, the most frequently used materials for such applications are thermoplastics, such as ABS, PLA, and composites. Through AM, intricate adjustments can be made to the size and shape of a prosthetic in order to generate optimum comfort, functionality, and aesthetic appeal (R. Kumar et al., 2018). Prosthetics can also be made more cheaply than with conventional methods of manufacturing, and so, they are more available to patients. AM is well-applied in dentistry for producing accurate crowns, bridges, or other dental restorations (Punia et al., 2023; Punia & Garg, 2024a, 2024b). SLA and DMLS are normally applied to materials like zirconia, cobalt-chrome, and resin-based photopolymers (Manero et al., 2019). These materials have good biocompatibility and mechanical properties as well as aesthetics, so that dental restorations not only can function efficiently but also can provide the esthetical requirements of restored teeth. AM helps in designing very accurate and personalized dental parts, which in turn increases both the contact between restorative and tooth tissue, as well as the durability of the restorations. Specialized orthopedics, including braces, splints, and supports, also receive additive manufacturing for fabrication (Kaushik et al., 2023). Thermoplastics, on one hand, include nylon and polypropylene, while elastomers can be printed using FDM as well as SLS materials. Ear pieces and hearing aids, which are popular for the earmolds, are today developed from 3D scanning and AM technologies (Rezaie et al., 2023; Tahayeri et al., 2018). Additive manufacturing also has application in designing the scaffold for tissue engineering in which the object produced by the additive manufacturing process is meant for growing new tissue such as skin, bone, or cartilage. These scaffolds offer the cells the architecture they need and allow budding cells to organize and develop into tissues that can be used for transplant or physiological repair (Pavan Kalyan & Kumar, 2022). Even though more of an experiment, bioprinting can construct fully functional tissues and organs. Scientists are developing the way to print organs such as livers, kidneys, and tissues for the heart with the intention to later on provide individual organ transplantations and cure ailments (Kumar & Malviya, 2024).

All the above applications show how implant manufacturing for medical use is the best since it involves uniqueness, precision, and speed. With the availability of AM technologies, namely, FDM, SLA, SLS, and DMLS, manufacturers are in a position to produce complex medical parts and even equipment, which are suitably designed to fit individual patients thus helping in the enhancement of efficiency of treatment and value addition to the patients. In medical AM, the materials include metals and polymers that are biocompatible, and the cells used in bioprinting make these parts not only functional but also safe for implantation and medical procedures (Figure 6.5).

6.3.3 Automotive sector

AM has caused waves in the automotive sector through conventional applications such as tooling and prototyping, lower costs of manufacturing, and the capacity for the production of intricate, weightless, and, customized products (Lim et al., 2016). Among the most frequently reported uses of AM technology in the automotive industry is the rapid prototyping of automotive parts. When it comes to components such as outside panels, dashboards and the operational parts of an engine, using FDM or SLA (Stereolithography) is acceptable to create prototypes. The materials used in these prototypes include ABS, PLA, and photopolymers

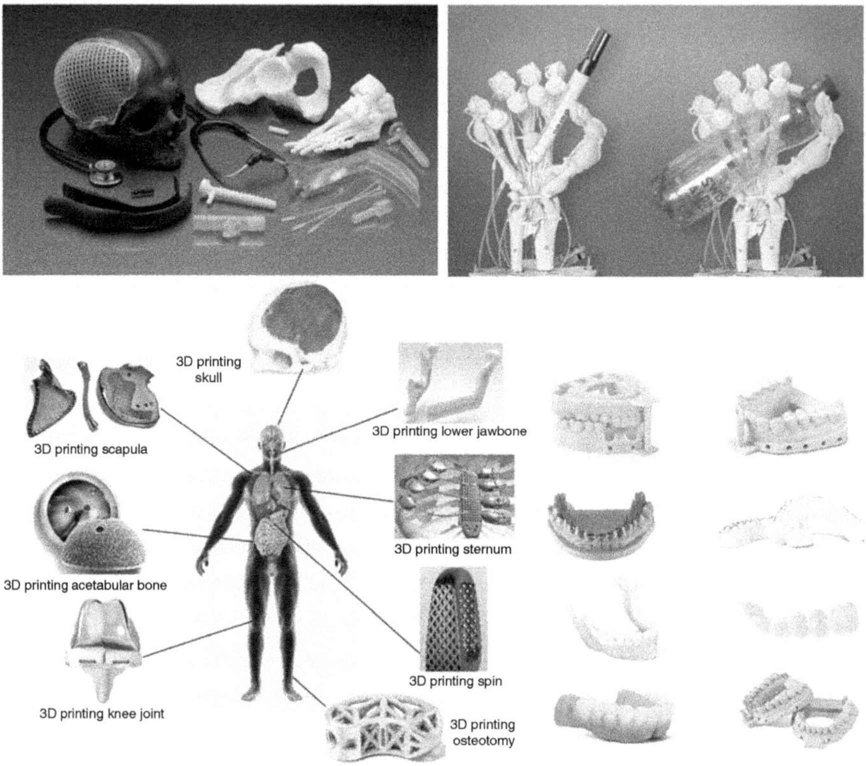

Figure 6.5 Application of AM in healthcare.

(Vasco, 2021). These prototypes enable the manufacturer to assess the form, fit, and function in a short span before laying out substantial capital in tooling and manufacturing. AM is also used for creating functional parts such as custom-made brackets, bolts, and clips. These parts are created out of thermoplastic elastomers, nylon, and polycarbonate by employing the FDM and the SLS methods (Yao & Ruan, 2024). These components need to have certain mechanical characteristics, which include high tensile strength and durability. AM is appropriate when used for a relatively small number of components or custom parts. Highly detailed parts like engine intake manifolds, fuel injectors, and even part of turbochargers are all ideal for additive manufacturing. The DMLS and SLM (selective laser melting) methods are used. Selective laser sintering has been more commonly used to manufacture one-off internal components inclusive of air vents, consoles, and seat items. These components are portable and can easily be adapted to suit the required scales; they enhance beauty and increased comfortability in cars, among other things (Ganesh Sarvankar & Yewale, 2019; Prabhu et al., 2023) (Figure 6.6).

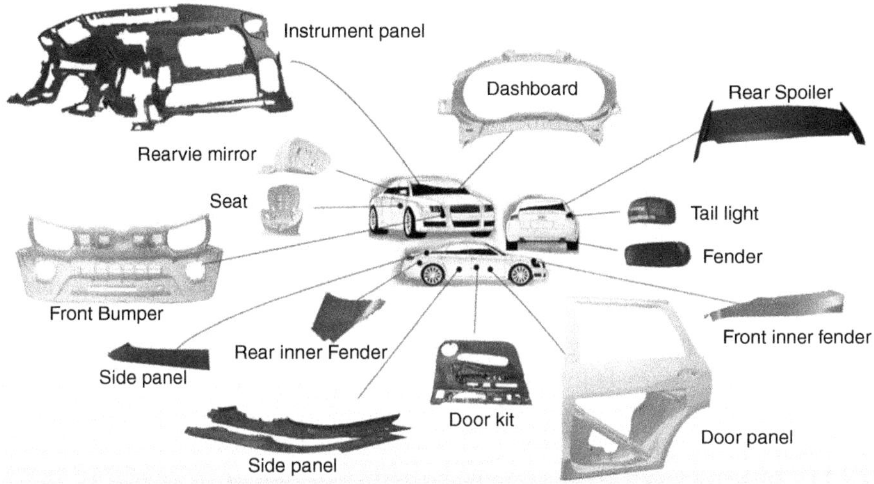

Figure 6.6 Application of AM in automotive sector.

Thus, in connection with the widespread use of cases, such as electric vehicles (EVs), additive manufacturing is used to create intricate battery parts. Some of the subassemblies, such as battery casings, cooling trays, and connectors are produced through SLS or DMLS technologies in AM. These include thermoplastics, aluminum, and copper alloys, which are conductive, lightweight, which are ideal for any EV battery systems (Coimbra et al., 2021).

To summarize, the use of additive manufacturing in the automotive industry means more freedom in design, shorter time to market, and the opportunity to manufacture singular unique pieces. Another advantage would be the fact that manufacturers are able to utilize different thermoplastic materials and high-performance metals to improve product performance, cut weight, and decrease costs, all in terms of meeting desired quality.

6.3.4 Educational tools

In the education sector today, AM is rapidly becoming a game-changer in how students study and learn, how teachers teach, and how educational assets within institutions are built and produced (Sun & Li, 2018). As a result, AM can be effectively utilized in creating educational information, materials, and models for better understanding of real tangible objects and for the purpose of use, which stimulates children's inventive minds and gives them the chance to improve their reasoning. The following is a list of several important educational applications of the use of additive manufacturing (Ford & Minshall, 2019).

Perhaps one of the most effective applications of AM in education involves creating models that would aid in teaching the complex science concepts. Such representations include, but are not restricted to, molecular models, anatomical models, geological models, or even planetary models. These models assist students in gaining better insight with regard to issues such as chemistry, biology, and geology among others due to their modelled display (Huang & Lin, 2017). In engineering and design classes, AM is employed to create models of mechanical, architectural, or electrical structures. These may be functional models of mechanical, civil structures, or a circuit board design. Unfortunately, to create the required products, students have to spend more time and money in traditional methods of manufacturing, but with additive manufacturing students are able to create prototypes and learn from their application. Education in art and design makes use of AM to design and develop one-of-a-kind pieces of sculpture and other forms of art. Students also take advantage of the fact that with 3D Printing, they can produce structures that cannot be designed using the conventional forms of anamorphic sculpture. This enables students to go further in their imaginative system and fuse art with technology (Menano et al., 2019; Schelly et al., 2015).

In conclusion, the EMS technology commonly referred to as additive manufacturing is revolutionising the education industry in terms of how students are taught and how tutors train their students. In the case of such applications as creating personalized models or assistive devices or making various engineering models and prototypes, AM facilitates the provision of quality and, in many cases, student-specific educational aids that are interactive. We must not underestimate versatile and inexpensive technologies like SLA, FDM, and SLS, along with available materials like PLA, ABS, resin, and TPU, which can be a valuable aid in schools, colleges, and research centers.

6.4 FUTURE DIRECTIONS OF AM

The future trends of additive manufacturing (AM) can be seen as prospective in many fields, including its materials, technologies, and uses. Some key areas are discussed in this section.

6.4.1 Advanced materials

The progress of AM greatly depends on the availability of new and improved materials. The current investigation is concentrated on new polymeric, metallic, ceramic, and composite materials that can improve the performances of the aforementioned basic properties compared to the traditional materials (Wang et al., 2024). Besides, including nanomaterials into AM processes as reinforcement like carbon nanotubes or graphene, can upgrade mechanical properties like strength-to-weight ratio, electrical conductivity, and thermal stability. These materials allow the manufacture of new components for the aerospace, automotive, and medical industries where conventional manufacturing approaches are restricted by material capabilities (Tuli et al., 2024).

6.4.2 Sustainability

The maturity of sustainability concept in AM has been researched widely because of efficient use of resources such as waste, energy, and material. AM generates inherently less waste than subtractive manufacturing because no material is wasted in building up a product through depositions (Liu et al., 2019). Moreover, AM power and sustainability goals is motivating recyclable and bio-based material development in the technology. For example, biopolymers and recycled feedstock can be employed to decrease the company's impact on the environment (Tyagi et al., 2024). Further, the inclusion of energy-efficient printing including the use of solar-powered 3D Printing or the use of energy-efficient modes during the printing process, is becoming crucial for attaining a sustainable manufacturing strategy (Al Rashid & Koç, 2023).

6.4.3 Multi-material printing

Another category is multi-material AM which allows the application of various materials at once, which means, creating heterophonic structures with the desirable mechanical characteristics at once. It also enables the creation of complex parts, where material is both the structural element and the electrical conductor, or where sensors are incorporated directly into the structure of the part, or where conductive pathways or structural reinforcements are required and further processes are unnecessary. Multi-material printing means that the materials can exhibit different mechanical, thermal, or electrical characteristics, which makes it possible to increase the variety of applications of the AM-created parts by adding various types of materials, especially in aerospace, automotive, and electronics industries (Hasanov et al., 2022; Roach et al., 2019).

6.4.4 Speed and scalability

It is vital for AM to scale up the rate and size for it to be suitable for industrial use. Recent advancements in printing technologies include the continuous liquid interface production (CLIP) and the high-speed sintering (HSS) to increase the printing speeds by a factor while keeping the quality of the parts produced constant or better. Further, in efforts to ramp up AM processes, it is concerned with the rate at which a machine can produce objects and the type of material to use for high volume production (Fang et al.,

2017; Walker et al., 2019). These enhancements could minimize the lead time and manufacturing cost – and thus bring AM closer to the conventional manufacturing for industries in need of short-run production requiring first iteration prototypes.

6.4.5 Automation and AI

AI and ML in AM are a revolutionary opportunity in term of streamlining the AM process, quality assurance, and maintenance service. Machine learning is capable of making adjustments to the build paths, the flow characteristics for supports and material, as well as related parameters in real-time, thereby benefiting the quality and time efficiency of the printing process. Besides, quality control systems aided by artificial intelligence can inform operators about defects and prevent its occurrence during the build process, which would improve the reliability of the AM-produced parts. ML also plays a central role in creating self-sustaining systems to address different conditions of deployment and increase reliability of the end products. Automation reduces time earlier spent on production and heavily diminishes human involvement in the process, increasing productivity, and cost-effectiveness in AM (Regassa Hunde & Debebe Woldeyohannes, 2022).

6.4.6 Integration with traditional manufacturing

When conventionally combined with other manufacturing techniques like CNC machining, AM-based hybrid manufacturing is steadily growing, since it leverages the advantages of the two techniques. AM is used to generate new designs and detailed geometric forms within less time when compared to conventional manufacturing techniques, which provide precision, surface texture, and the type of material to be used. This low-to-medium risk strategy allows the creation of parts with high performance requirements, such as high tolerance and quality, while also offering design freedom and rapid manufacturing values of AM (Balletti et al., 2017). The infusion of AM with already-existing industrial production platforms is ideal for industries targeting to exploit the strength of both technologies, especially in aerospace, automotive, and medical sectors.

6.5 CONCLUSION

Additive manufacturing is now widely recognized as an integral part of modern manufacturing with peculiar advantages of design freedom, customization, and sustainability. That is why AM is transforming industries such as aerospace, healthcare, automotive, and construction, offering the possibility to create geometries, lightweight structures, and application-specific components that they could not make before. The classification of

AM technologies shows how each of them is designed to accommodate particular application and materials usage. The flexibility of AM also guarantees the continued applicability of AM in a wide range of industrial and consumer requirements. Therefore, progress in the area of materials, processes, and incorporation of multiple materials in AM will increasingly drive the technology's industrial applicability and cost-effectiveness. It is believed that integration with other manufacturing methods, increased automation, and the desire to use biocompatible and environment friendly materials will expand its applications. With industries embracing AM, innovation, waste reductions, and localized manufacturers will strengthen the roles of AM as an enabler of the fourth industrial revolution. Lastly, this article shows that AM has the potential to transform manufacturing industries and outlines how it will evolve in manufacturing and other sectors.

REFERENCES

Al Rashid, A., Ahmed, W., Khalid, M. Y., & Koç, M. (2021). Vat photopolymerization of polymers and polymer composites: Processes and applications. *Additive Manufacturing*, 47. https://doi.org/10.1016/j.addma.2021.102279

Al Rashid, A., & Koç, M. (2023). Additive manufacturing for sustainability and circular economy: Needs, challenges, and opportunities for 3D printing of recycled polymeric waste. *Materials Today Sustainability*, 24, 100529. https://doi.org/10.1016/j.mtsust.2023.100529

Altıparmak, S. C., Yardley, V. A., Shi, Z., & Lin, J. (2022). Extrusion-based additive manufacturing technologies: State of the art and future perspectives. *Journal of Manufacturing Processes*, 83, 607–636. https://doi.org/10.1016/j.jmapro.2022.09.032

Angrish, A. (2014). A critical analysis of additive manufacturing technologies for aerospace applications. *IEEE Aerospace Conference Proceedings*, 1–6. https://doi.org/10.1109/AERO.2014.6836456

Balletti, C., Ballarin, M., & Guerra, F. (2017). 3D printing: State of the art and future perspectives. *Journal of Cultural Heritage*, 26, 172–182.

Bandyopadhyay, A., & Heer, B. (2018). Additive manufacturing of multi-material structures. *Materials Science and Engineering: R: Reports*, 129, 1–16.

Bhatia, A., & Sehgal, A. K. (2023). Additive manufacturing materials, methods and applications: A review. *Materials Today: Proceedings*, 81, 1060–1067.

Bikas, H., Stavropoulos, P., & Chryssolouris, G. (2016). Additive manufacturing methods and modeling approaches: A critical review. *International Journal of Advanced Manufacturing Technology*, 83(1–4), 389–405. https://doi.org/10.1007/s00170-015-7576-2

Coimbra, M. R. C., Barbosa, T. P., & Vasques, C. M. A. (2021). Preliminary design and validation of a 3D-printed continuously variable transmission for an electric vehicle prototype †. *Engineering Proceedings*, 11(1), 84. https://doi.org/10.3390/ASEC2021-11178

Davoudinejad, A. (2021). Vat photopolymerization methods in additive manufacturing. In *Additive Manufacturing* (pp. 159–181). Elsevier. https://doi.org/10.1016/B978-0-12-818411-0.00007-0

Delgado Camacho, D., Clayton, P., O'Brien, W. J., Seepersad, C., Juenger, M., Ferron, R., & Salamone, S. (2018). Applications of additive manufacturing in the construction industry – A forward-looking review. *Automation in Construction*, 89, 110–119. https://doi.org/10.1016/j.autcon.2017.12.031

Dodziuk, H. (2016). Applications of 3D printing in healthcare. *Kardiochirurgia i Torakochirurgia Polska*, 13(3), 283–293. https://doi.org/10.5114/kitp.2016.62625

Fang, F., Aabith, S., Homer-Vanniasinkam, S., & Tiwari, M. K. (2017). High-resolution 3D printing for healthcare underpinned by small-scale fluidics. In *3D Printing in Medicine* (pp. 167–206). Elsevier. https://doi.org/10.1016/B978-0-08-100717-4.00023-5

Ford, S., & Minshall, T. (2019). Invited review article: Where and how 3D printing is used in teaching and education. *Additive Manufacturing*, 25, 131–150. https://doi.org/10.1016/j.addma.2018.10.028

Ganesh Sarvankar, S., & Yewale, S. N. (2019). Additive manufacturing in automobile industry. *International Journal of Research in Aeronautical and Mechanical Engineering*, 7(4), 1–10.

Gui, Y., Aoyagi, K., Bian, H., & Chiba, A. (2022). Detection, classification and prediction of internal defects from surface morphology data of metal parts fabricated by powder bed fusion type additive manufacturing using an electron beam. *Additive Manufacturing*, 54, 102736. https://doi.org/10.1016/j.addma.2022.102736

Hasanov, S., Alkunte, S., Rajeshirke, M., Gupta, A., Huseynov, O., Fidan, I., Alifui-Segbaya, F., & Rennie, A. (2022). Review on additive manufacturing of multi-material parts: Progress and challenges. *Journal of Manufacturing and Materials Processing*, 6(1), 4. https://doi.org/10.3390/jmmp6010004

Huang, T. C., & Lin, C. Y. (2017). From 3D modeling to 3D printing: Development of a differentiated spatial ability teaching model. *Telematics and Informatics*, 34(2), 604–613. https://doi.org/10.1016/j.tele.2016.10.005

Jadhav, A., & Jadhav, V. S. (2022). A review on 3D printing: An additive manufacturing technology. *Materials Today: Proceedings*, 62, 2094–2099. https://doi.org/10.1016/j.matpr.2022.02.558

Javaid, M., & Haleem, A. (2019). Current status and applications of additive manufacturing in dentistry: A literature-based review. *Journal of Oral Biology and Craniofacial Research*, 9(3), 179–185. https://doi.org/10.1016/j.jobcr.2019.04.004

Javaid, M., Haleem, A., Singh, R. P., & Suman, R. (2022). 3D printing applications for healthcare research and development. *Global Health Journal*, 6(4), 217–226. https://doi.org/10.1016/j.glohj.2022.11.001

Jiménez, M., Romero, L., Domínguez, I. A., Espinosa, M. D. M., & Domínguez, M. (2019). Additive manufacturing technologies: An overview about 3D printing methods and future prospects. *Complexity*, 2019. https://doi.org/10.1155/2019/9656938

Joshi, S. C., & Sheikh, A. A. (2015). 3D printing in aerospace and its long-term sustainability. *Virtual and Physical Prototyping*, 10(4), 175–185. https://doi.org/10.1080/17452759.2015.1111519

Kalender, M., Kilic, S. E., Ersoy, S., Bozkurt, Y., & Salman, S. (2019). Additive manufacturing and 3D printer technology in aerospace industry. *Proceedings of 9th International Conference on Recent Advances in Space Technologies, RAST 2019*, 689–695. https://doi.org/10.1109/RAST.2019.8767881

Kaushik, A., Punia, U., Gahletia, S., Garg, R. K., & Chhabra, D. (2023). Identification and overcoming key challenges in the 3D printing revolution. *3D Printing and Sustainable Product Development*, 87–104. https://doi.org/10.1201/9781003306238-5

Kumar, R., Singh, R., & Farina, I. (2018). On the 3D printing of recycled ABS, PLA and HIPS thermoplastics for structural applications. *PSU Research Review*, 2(2), 115–137. https://doi.org/10.1108/PRR-07-2018-0018

Kumar, S., & Malviya, R. (2024). Bioprinting of hepatic tissue using 3D technology: Transitioning beyond laboratory models to real-world applications in medical treatments. *Applied Materials Today*, 39, 102307. https://doi.org/10.1016/j.apmt.2024.102307

Kunovjanek, M., Knofius, N., & Reiner, G. (2022). Additive manufacturing and supply chains–a systematic review. *Production Planning and Control*, 33(13), 1231–1251. https://doi.org/10.1080/09537287.2020.1857874

Lakhdar, Y., Tuck, C., Binner, J., Terry, A., & Goodridge, R. (2021). Additive manufacturing of advanced ceramic materials. *Progress in Materials Science*, 116, 100736. https://doi.org/10.1016/j.pmatsci.2020.100736

Lim, C. W. J., Le, K. Q., Lu, Q., & Wong, C. H. (2016). An overview of 3-D printing in manufacturing, aerospace, and automotive industries. *IEEE Potentials*, 35(4), 18–22. https://doi.org/10.1109/MPOT.2016.2540098

Liu, J., Sun, L., Xu, W., Wang, Q., Yu, S., & Sun, J. (2019). Current advances and future perspectives of 3D printing natural-derived biopolymers. *Carbohydrate Polymers*, 207, 297–316. https://doi.org/10.1016/j.carbpol.2018.11.077

Mahamood, R. M., Jen, T. C., Akinlabi, S. A., Hassan, S., Abdulrahman, K. O., & Akinlabi, E. T. (2021). Role of additive manufacturing in the era of Industry 4.0. *Additive Manufacturing: A Tool for Industrial Revolution 4.0*, 11, 107–126. https://doi.org/10.1016/B978-0-12-822056-6.00003-5

Manero, A., Smith, P., Sparkman, J., Dombrowski, M., Courbin, D., Kester, A., Womack, I., & Chi, A. (2019). Implementation of 3D printing technology in the field of prosthetics: Past, present, and future. *International Journal of Environmental Research and Public Health*, 16(9), 1641. https://doi.org/10.3390/ijerph16091641

Martinez, D. W., Espino, M., Cascolan, H. M., Crisostomo, J. L., & Dizon, J. R. (2022). A comprehensive review on the application of 3D printing in the aerospace industry. *Key Engineering Materials*, 913, 27–34. https://doi.org/10.4028/p-94a9zb

Mehrpouya, M., Dehghanghadikolaei, A., Fotovvati, B., Vosooghnia, A., Emamian, S. S., & Gisario, A. (2019). The potential of additive manufacturing in the smart factory industrial 4.0: A review. *Applied Sciences*, 9(18), 3865. https://doi.org/10.3390/app9183865

Menano, L., Fidalgo, P., Santos, I. M., & Thormann, J. (2019). Integration of 3D printing in art education: A multidisciplinary approach. *Computers in the Schools*, 36(3), 222–236. https://doi.org/10.1080/07380569.2019.1643442

Meng, M., Wang, J., Huang, H., Liu, X., Zhang, J., & Li, Z. (2023). 3D printing metal implants in orthopedic surgery: Methods, applications and future prospects. *Journal of Orthopaedic Translation*, 42, 94–112. https://doi.org/10.1016/j.jot.2023.08.004

Moeinfar, K., Khodabakhshi, F., Kashani-bozorg, S. F., Mohammadi, M., & Gerlich, A. P. (2022). A review on metallurgical aspects of laser additive manufacturing (LAM): Stainless steels, nickel superalloys, and titanium alloys. *Journal of Materials Research and Technology*, 16, 1029–1068. https://doi.org/10.1016/j.jmrt.2021.12.039

Motyl, B., & Filippi, S. (2021). Trends in engineering education for additive manufacturing in the industry 4.0 era: A systematic literature review. *International Journal on Interactive Design and Manufacturing*, 15(1), 103–106. https://doi.org/10.1007/s12008-020-00733-1

Mouzakis, D. E. (2018). Advanced technologies in manufacturing 3D-layered structures for defense and aerospace. *Lamination - Theory and Application*, 74331. https://doi.org/10.5772/intechopen.74331

Mpofu, T. P., Mawere, C., & Mukosera, M. (2014). The impact and application of 3D printing technology. *International Journal of Science and Research (IJSR)*, 3(6), 2148–2152. https://www.academia.edu/download/34056587/MDIwMTQ2NzU_.pdf%0Ahttps://www.researchgate.net/publication/291975129

Murr, L. E. (2020). Global trends in the development of complex, personalized, biomedical, surgical implant devices using 3D printing/additive manufacturing: A review. *Medical Devices & Sensors*, 3(6), e10126. https://doi.org/10.1002/mds3.10126

Najmon, J. C., Raeisi, S., & Tovar, A. (2019). Review of additive manufacturing technologies and applications in the aerospace industry. In F. Froes, R. Boyer, & J. H. Rao (Eds.), *Additive Manufacturing for the Aerospace Industry* (pp. 7–31). Elsevier. https://doi.org/10.1016/B978-0-12-814062-8.00002-9

Oleff, A., Küster, B., Stonis, M., & Overmeyer, L. (2021). Process monitoring for material extrusion additive manufacturing: A state-of-the-art review. *Progress in Additive Manufacturing*, 6(4), 705–730. https://doi.org/10.1007/s40964-021-00192-4

Pavan Kalyan, B., & Kumar, L. (2022). 3D printing: Applications in tissue engineering, medical devices, and drug delivery. *AAPS PharmSciTech*, 23(4), 92. https://doi.org/10.1208/s12249-022-02242-8

Pazhamannil, R. V., & Govindan, P. (2020). Current state and future scope of additive manufacturing technologies via vat photopolymerization. *Materials Today: Proceedings*, 43, 130–136. https://doi.org/10.1016/j.matpr.2020.11.225

Pereira, T., Kennedy, J. V., & Potgieter, J. (2019). A comparison of traditional manufacturing vs additive manufacturing, the best method for the job. *Procedia Manufacturing*, 30, 11–18. https://doi.org/10.1016/j.promfg.2019.02.003

Picard, M., Mohanty, A. K., & Misra, M. (2020). Recent advances in additive manufacturing of engineering thermoplastics: Challenges and opportunities. *RSC Advances*, 10(59), 36058–36089. https://doi.org/10.1039/d0ra04857g

Prabhu, L., Krishnamoorthi, S., Joseph, J., George, J. K., & Naveen, P. (2023). Development of advanced 3D printing technology in automotive industry. *AIP Conference Proceedings*, 2523(1). https://doi.org/10.1063/5.0110513

Punia, U., & Garg, R. K. (2024a). Enhancing mechanical performance of 3D printable PMMA resin through strategic incorporation of SS 316L nanoparticles for dental applications. *International Journal on Interactive Design and Manufacturing*. https://doi.org/10.1007/s12008-024-02036-1

Punia, U., & Garg, R. K. (2024b). Reinforcing dental restoration material PMMA with SS 316L nanoparticles for mechanical performance analysis. *Progress in Additive Manufacturing*. https://doi.org/10.1007/s40964-024-00800-z

Punia, U., Kaushik, A., Garg, R. K., Sharma, A., & Chhabra, D. (2023). Additive manufacturing requirements for dental implementation. *3D Printing and Sustainable Product Development*, 151–171. https://doi.org/10.1201/9781003306238-9

Regassa Hunde, B., & Debebe Woldeyohannes, A. (2022). Future prospects of computer-aided design (CAD) – A review from the perspective of artificial intelligence (AI), extended reality, and 3D printing. *Results in Engineering*, 14, 100478. https://doi.org/10.1016/j.rineng.2022.100478

Rezaie, F., Farshbaf, M., Dahri, M., Masjedi, M., Maleki, R., Amini, F., Wirth, J., Moharamzadeh, K., Weber, F. E., & Tayebi, L. (2023). 3D printing of dental prostheses: Current and emerging applications. *Journal of Composites Science*, 7(2), 80. https://doi.org/10.3390/jcs7020080

Roach, D. J., Hamel, C. M., Dunn, C. K., Johnson, M. V., Kuang, X., & Qi, H. J. (2019). The m4 3D printer: A multi-material multi-method additive manufacturing platform for future 3D printed structures. *Additive Manufacturing*, 29, 100819. https://doi.org/10.1016/j.addma.2019.100819

Rojek, I., Mikołajewski, D., Dostatni, E., & Macko, M. (2020). Ai-optimized technological aspects of the material used in 3D printing processes for selected medical applications. *Materials*, 13(23), 1–19. https://doi.org/10.3390/ma13235437

Schelly, C., Anzalone, G., Wijnen, B., & Pearce, J. M. (2015). Open-source 3-D printing technologies for education: Bringing additive manufacturing to the classroom. *Journal of Visual Languages and Computing*, 28, 226–237. https://doi.org/10.1016/j.jvlc.2015.01.004

Scime, L., & Beuth, J. (2018). A multi-scale convolutional neural network for autonomous anomaly detection and classification in a laser powder bed fusion additive manufacturing process. *Additive Manufacturing*, 24, 273–286. https://doi.org/10.1016/j.addma.2018.09.034

Shaikh, S., Nahar, P., & Ali, H. M. (2021). Current perspectives of 3D printing in dental applications. *Brazilian Dental Science*, 24(3), 1–15. https://doi.org/10.14295/bds.2021.v24i3.2481

Shanmugam, R., Ramoni, M. O., Chandran, J., Mohanavel, V., & Pugazhendhi, L. (2021). A review on the significant classification of additive manufacturing. *Journal of Physics: Conference Series*, 2027(1), 12026. https://doi.org/10.1088/1742-6596/2027/1/012026

Singh, A. B. (2024). Transforming healthcare: A review of additive manufacturing applications in the healthcare sector. *Engineering Proceedings*, 72(1), 2. https://doi.org/10.3390/engproc2024072002

Singh, R., Gupta, A., Tripathi, O., Srivastava, S., Singh, B., Awasthi, A., Rajput, S. K., Sonia, P., Singhal, P., & Saxena, K. K. (2019). Powder bed fusion process in additive manufacturing: An overview. *Materials Today: Proceedings*, 26, 3058–3070. https://doi.org/10.1016/j.matpr.2020.02.635

Sun, Y., & Li, Q. (2018). The application of 3D printing in STEM education. *Proceedings of 4th IEEE International Conference on Applied System Innovation 2018, ICASI 2018*, 20(1), 1115–1118. https://doi.org/10.1109/ICASI.2018.8394476

Tahayeri, A., Morgan, M. C., Fugolin, A. P., Bompolaki, D., Athirasala, A., Pfeifer, C. S., Ferracane, J. L., & Bertassoni, L. E. (2018). 3D printed versus conventionally cured provisional crown and bridge dental materials. *Dental Materials*, 34(2), 192–200. https://doi.org/10.1016/j.dental.2017.10.003

Tan, C., Li, R., Su, J., Du, D., Du, Y., Attard, B., Chew, Y., Zhang, H., Lavernia, E. J., Fautrelle, Y., Teng, J., & Dong, A. (2023). Review on field assisted metal additive manufacturing. *International Journal of Machine Tools and Manufacture*, 189, 104032. https://doi.org/10.1016/j.ijmachtools.2023.104032

Tofail, S. A. M., Koumoulos, E. P., Bandyopadhyay, A., Bose, S., O'Donoghue, L., & Charitidis, C. (2018). Additive manufacturing: scientific and technological challenges, market uptake and opportunities. *Materials Today*, 21(1), 22–37. https://doi.org/10.1016/j.mattod.2017.07.001

Tuli, N. T., Khatun, S., & Rashid, A. Bin. (2024). Unlocking the future of precision manufacturing: A comprehensive exploration of 3D printing with fiber-reinforced composites in aerospace, automotive, medical, and consumer industries. *Heliyon*, 10(5). https://doi.org/10.1016/j.heliyon.2024.e27328

Tyagi, A., Punia, U., Dadhich, A., & Meena, S. L. (2024). Global perspective of natural fibre reinforced composites: Properties, and applications. *Journal of The Institution of Engineers (India): Series C*. https://doi.org/10.1007/s40032-024-01076-6

Vasco, J. C. (2021). Additive manufacturing for the automotive industry. In *Additive Manufacturing* (pp. 505–530). Elsevier. https://doi.org/10.1016/B978-0-12-818411-0.00010-0

Walker, D. A., Hedrick, J. L., & Mirkin, C. A. (2019). Rapid, large-volume, thermally controlled 3D printing using a mobile liquid interface. *Science*, 366(6463), 360–364. https://doi.org/10.1126/science.aax1562

Wang, Y., Wu, T., & Huang, G. (2024). State-of-the-art research progress and challenge of the printing techniques, potential applications for advanced ceramic materials 3D printing. *Materials Today Communications*, 40, 110001. https://doi.org/10.1016/j.mtcomm.2024.110001

Wawryniuk, Z., Brancewicz-Steinmetz, E., & Sawicki, J. (2024). Revolutionizing transportation: an overview of 3D printing in aviation, automotive, and space industries. *International Journal of Advanced Manufacturing Technology*, 134(7–8), 3083–3105. https://doi.org/10.1007/s00170-024-14226-y

Yao, W., & Ruan, Y. (2024). Development and application of 3D printing manufacturing technology in intelligent vehicle design. *International Journal of Uncertainty, Fuzziness and Knowledge-Based Systems*, 32(4), 491–513. https://doi.org/10.1142/S0218488524400051

Chapter 7

Materials in additive manufacturing

Current trends and future directions

Deepak Dahiya, Kiran Kaushik, Upender Punia, Ankit Tyagi, and Anant Bhardwaj

ABBREVIATIONS

AM	additive manufacturing
AI	artificial intelligence
ML	machine learning
3D	three dimensional

7.1 INTRODUCTION

Three Dimensional (3D) printing, or Additive manufacturing (AM) is another emerging technology that has transformed the design and manufacturing of intricate structures in various fields, including aviation and medical science. This chapter is a systematic account of different materials used in AM, that include polymers, metals, ceramics, composites, biomaterials, and nanomaterials. Main concerns include material characteristics, processes they are compatible with, and developments that have led to the adoption of certain materials in various uses. There is a special focus on the coupling between material choice and advances in AM techniques such as multi-material printing and FGMs. Moreover, challenges related to sustainability issues are provided, material recycling, and seeking for eco-friendly materials instead of conventional precursors. Lastly, the future directions section is a discussion of advances in smart materials and bio-inspired systems, as well as advances in applying artificial intelligence in material selection and the optimization of the next-generation AM processes. This work is a useful base for researchers, engineers, and supporters, who want to explore the material foundation of additive manufacturing (Figure 7.1).

Firstly, create digital model or 3D file by drawing with the help of computer-aided design (CAD) software. Subsequently, the concept designed digitally is translated into the STL file format. The machine then scans and begins to build the object in a layer-by-layer queue. Then there is the process of product realization done as well. Ironically, AM is more than the machines; it has a great deal to do with the employed material. All of these

DOI: 10.1201/9781003607052-7

Figure 7.1 Flow chart showing the manufacturing process.

Figure 7.2 Material used in additive manufacturing.

materials offer characteristics that define their suitability at various AM processes. For example, materials like PLA and ABS are used for creating models, strong metals like titanium and aluminum are used in aerospace and medical fields (Ngo et al., 2018). It is not just about seeking a material that will work: it is about seeking the material which will function efficiently in the cheapest manner while also being environment friendly. For instance, polymers that are bio-compatible have allowed for creation of one–of–a–kind medical implants and advanced alloys that have made lighter and far more efficient aircrafts. The research in this field is now turning to areas such as high-performance fiber-reinforced polymer matrix composites, biodegradable polymers, and self-healing polymers (Tyagi et al., 2024). Some of these developments are challenging the limits of AM and turning what had earlier been only a futuristic concept into a real possibility (Figure 7.2).

In this chapter, we will explore the different types of materials used in AM, how they interact with various printing processes, and the latest trends shaping the future of materials in this rapidly evolving field (Vaezi et al., 2013).

7.2 POLYMERS IN ADDITIVE MANUFACTURING

7.2.1 Polymers in additive manufacturing

Polymers are among the most widely used materials in additive manufacturing (AM) due to their versatility, ease of processing, and range of mechanical properties. They are broadly classified into thermoplastics, thermosets, and advanced polymers based on their molecular structure and thermal behavior (Gibson et al., 2015).

Polymer materials are among the largest used materials in AM because of their versatility, ease of processing, and various mechanical properties. They are categorized largely under thermoplastic, thermosetting and high-performance polymers by structural arrangement and its thermal properties (Gibson et al., 2015).

7.2.1.1 Thermoplastics

Thermoplastics are those material which melts when exposed to heat and hardens at room temperature, and are particularly suitable for AM methods such as FDM and SLS (Gibson et al., 2015). Polylactic acid (PLA) is an aliphatic thermoplastic derived from renewable source such as corn starch. This material is selected for its printability, low glass transition temperature, and its tendency to warp very little. For instance, PLA is employed in hobbies where enthusiasts develop unique varieties of figurines, phone cases, and even home accessories such as key racks (Jones et al., 2020). Acrylonitrile Butadiene Styrene (ABS), which is quite strong, tough and has high impact strength, is ideally suited for use in products that are required in industries. But the heat is set at a higher range for printing and need heated bed to avoid warping. For ABS, one must think of the material in terms of what lies behind such things as toys that eschew the LEGO name but are made of ABS plastic, as well as automotive interior trims, among others. PETG stands for Polyethylene Terephthalate Glycol and it adds the strength of ABS and the ease of print of PLA. This material is chemically stable, fairly robust, and is applied extensively in molding functions parts, medical equipment, and food packages. An example is custom water bottles or protective face shields that may be used in a health care institution (Figure 7.3).

7.2.1.2 Thermosets

Thermosets are materials which undergo chemical reactions that make them change from a liquid or a soft solid to a solid state when exposed to heat, light, or a chemical reagent. For this reason, they are well-suited for applications where relatively high strength and resistance to chemical action are required (Gibson et al., 2015). The epoxy resins are widely used in AM processes such as vat photopolymerization also known as stereolithography because of their

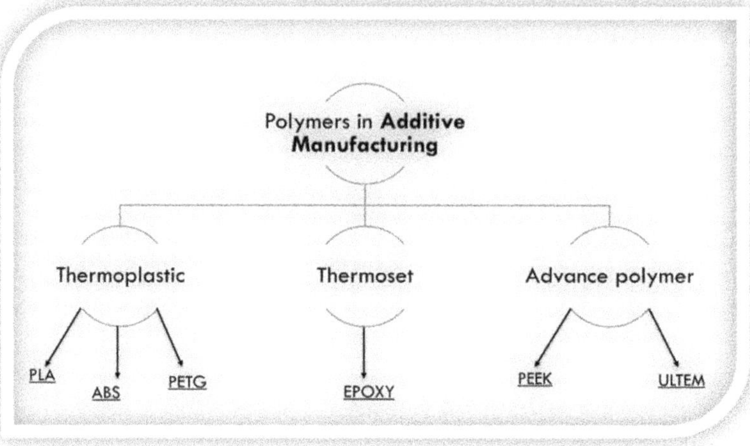

Figure 7.3 Types of polymers in additive manufacturing.

outstanding characteristics such as adhesion, tensile strength, and chemical resistance. They are essential in forming complicated intricate elements of high data resolution used in manufacture units like aerospace and automotive industries. Dedicated polymers have enhanced mechanical, thermal, and chemical performances that make them appropriate to be used in challenging industrial practices (Gibson et al., 2015). Polyetheretherketone, commonly referred to as PEEK, is a polyaryletter ether ketone that is characterized with high thermal stability, tensile strength, and chemical resistant responses. It is applied in aerospace, automotive, and in the construction of medical implants (Brown et al., 2022). ULTEM polyether sulfone resins is characterized by high heat stability, flame resistance, and high mechanical strength. This type of epoxy is mainly used in aerospace or electronic industries where low damping and high mechanical strength, as well as thermal properties, are required (Jackson & Green, 2020).

7.3 CERAMICS IN ADDITIVE MANUFACTURING

Advanced ceramic materials are preferred in AM because of their high-temperature stability, wear resistance, and chemical inertness. They are very important especially in parts that need to be reliable in harsh environments (Gibson et al., 2015). Various types of ceramics are oxides, carbides, nitrides, bio-ceramics, and glass. So, it indicates that oxide ceramics possess high heat resistance and corrosion properties. Aluminum Oxide (Al_2O_3) or alumina is another important ceramic, which offers its applications in wear-resistant parts such as cutting tools, electrical insulators, or in bio-medical applications like implants. For instance, ring-shaped and plate-shaped alumina components produced using 3D printers are incorporated into high-performance electronic appliances. Zirconium Oxide (ZrO_2) is highly desired due to the favorable characteristics of high hardness, and brittle resistance, commonly used in/on dental implants and oxygen-sensing applications (Brown et al., 2022). Dental crowns with AM offer patient's fitting restorations from AM components regarding their perspectives, weaving, morphology, parallelism, and occlusal planes. Since carbides and nitrides forms highly hard materials that can also endure high heat, they have been used in production of various components used in industries (Singh et al., 2018). Such use features high thermal conductivity and wear resistivity which are admirable for aerospace activities or engine parts (Jackson & Green, 2020). Rocket engines provide a practical example of 3D-printed nozzles made of silicon carbide known as SiC. Electronics is one of the common areas where BN is applied due to its good thermal conductivity while at the same time conducting electricity poorly. It is used in creating parts, such as heat sinks and insulating casings.

Ceramics are utilized in bio-medical applications because they are bio-ceramics while glass ceramics possess desired optical and mechanical properties (Gibson et al., 2015). Because of its similarity to the composition of natural bone tissue, hydroxyapatite (HA) is perfect for use in implants and prostheses, primarily in operations replacing bone tissues (Jones et al., 2020). These ceramic materials are used in dental restorations, glasses, space telescope mirrors among others. For instance, 3D-printed glass ceramic lenses enhance resolution in scientific equipment (Table 7.1).

Table 7.1 Ceramics in additive manufacturing

Category	*Type*	*Key properties*	*Applications*	*Examples*
Oxides	Aluminum oxide (Al_2O_3)	Thermal stability, corrosion resistance	Wear-resistant tools, insulators, implants	3D-printed alumina in electronics
	Zirconium oxide (ZrO_2)	Toughness, crack resistance	Dental implants, oxygen sensors	Custom-fit dental crowns
Carbides and nitrides	Silicon carbide (SiC)	Thermal conductivity, wear resistance	Aerospace components, engine parts	3D-printed SiC nozzles for rockets
	Boron nitride (BN)	Thermal conductivity, electrical insulation	Electronics, heat sinks	Insulating casings in electronics
Bio-ceramics and glass ceramics	Hydroxyapatite (HA)	Biocompatibility, bone-like structure	Bone implants, prosthetics	3D-printed HA scaffolds for surgery
	Glass ceramics	Optical and mechanical properties	Dental restorations, optical lenses	3D-printed lenses in scientific instruments

7.4 COMPOSITES IN ADDITIVE MANUFACTURING

These composite materials have really enhanced the capabilities of AM by allowing for a material with properties that are superior to the two base materials. Being made of two or more dissimilar materials that are bonded together, composites exhibit improved mechanical, thermal, and chemical properties, or performance making them a must-use material in many industries. This section presents information on various forms of composites, the use of each form, and the importance of the characteristics by formulating an essential part of AM technologies today (Figure 7.4).

7.4.1 Polymer-matrix composites (PMC)

Polymer matrix composites are one of the key pillars of AM as described by incorporating reinforcing fibers with a polymer matrix that can be glass, carbon, or metals (Punia & Garg, 2024a, 2024b). These composites are admired due to their flexibility, fabrication convenience, and low cost. The polymer matrix has the primary responsibility of positioning the reinforcing fibers and acting as a shield against environmental conditions (Gibson et al., 2015). PMC's features make them useful in aerospace, automobiles, sports gear, and construction industries because of their high strength-to-weight ratios. For instance, lightweight in aerospace structures such as layup for an airplane and high-reach small sporting equipment require PMCs that must

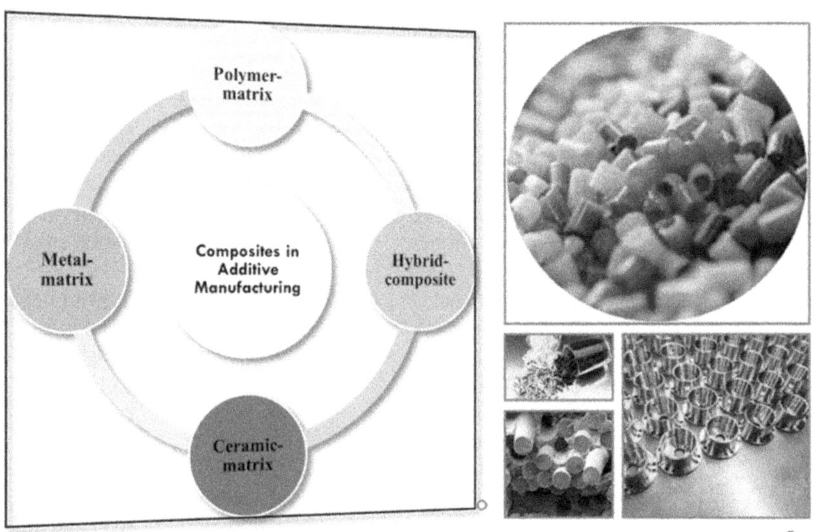

Figure 7.4 Composites in additive manufacturing.

meet specific performance standards. But they may possess lower thermal stability and toughness compared to their metal-matrix alloy composites (Singh et al., 2019).

7.4.2 Metal-matrix composites (MMC)

In metal-matrix composites, a metal acts as the base or matrix material which can be reinforced with ceramic, carbon fibers, or metal particles. This enhances the inherent properties of metals with the synergies of the reinforcement materials, as shown by Chawla et al. (2021). The key customers of MMCs are industries that need materials with high strength, heat, and wear resistance such as aerospace, automotive, and defense industries. Examples include parts of jet engine and automotive braking system (Jackson & Green, 2020). These composites have enhanced characteristics such as mechanical strength and modulus of elasticity and thermal conductivity. Nevertheless, their preparation could be more challenging, and they may be denser than polymer-matrix composites.

7.4.3 Ceramic-matrix composites (CMC)

Ceramic-matrix composites overcome the brittleness and fracture issues of traditional ceramics by incorporating reinforcing fibers, typically carbon or other ceramics. These composites are ideal for high-temperature applications (Gibson et al., 2015).

CMCs are used in turbine engines, aerospace thermal protection systems, and cutting tools. Their ability to endure extreme temperatures and harsh environments makes them indispensable in these fields CMCs offer high thermal resistance, strength, and wear resistance. However, they remain more brittle compared to metal-matrix composites (Jackson & Green, 2020).

7.4.4 Hybrid composites and emerging materials

The hybrid composites are those whose matrix materials or reinforcement phases consist of at least two different types. Other materials including graphene-based composites/nanocomposites are also emerging important application areas in AM because of their enhanced functionalities. These materials, which can possess desirable attributes such as strength, weight, and durability, can be used for current engineering applications. This combined way offers enhanced impact strength, thermal stability and the designed mechanical characteristics. Other material attributes include, electrical conductivity, and tensile strength which are also improved by newly developed composites like graphene-based composites (Table 7.2).

Table 7.2 Composites in additive manufacturing

Composite type	*Applications*	*Properties*	*Examples*
Polymer-matrix composites (PMC)	Aerospace, automotive, sports equipment, and construction industries.	Excellent corrosion resistance, low density, and flexibility.	Carbon fiber-reinforced polymer (CFRP)glassfiber-reinforced polymer (GFRP).
	For example, lightweight components in airplanes and high-performance sports gear (Jones & Taylor, 2020).	May have lower thermal stability and strength compared to MMCs (Singh et al., 2019).	
Metal-matrix composites (MMC)	Aerospace, automotive, and defense industries (e.g., jet engine parts).	Superior strength, stiffness, and thermal conductivity.	Aluminum matrix composites, titanium matrix composites, magnesium matrix composites.
	Examples include automotive brake systems (Jackson & Green, 2020).	Fabrication complexity and higher weight compared to PMCs (	
Ceramic-matrix composites (CMC)	Turbine engines, aerospace thermal protection systems, and cutting tools.	High thermal resistance, strength, and wear resistance.	Silicon carbide (SiC)-based composites, carbon-carbon composites.
	Ideal for extreme environments (Brown et al., 2022).	More brittle compared to MMCs (Jackson & Green, 2020).	
Hybrid composites and emerging materials	Automotive parts, wind turbine blades, advanced construction materials.	Balance of strength, weight, and durability.	Hybrid carbon/glass fiber composites, metal-polymer hybrids, graphene-based composites.
	Examples include graphene-based composites for electrical conductivity (Singh et al., 2019).	Improved impact resistance and thermal stability (Jones & Taylor, 2020).	

7.5 THE FUTURE OF MATERIALS FOR ADDITIVE MANUFACTURING (AM)

The further directions of both raw materials and products of additive manufacturing are becoming increasingly complex, smart, and personalized. This ranges from innovative smart materials, 4D Printing, and latest bio-printing materials that focuses mostly on performance, sustainability, and functionality among others (Kaushik et al., 2023; Punia et al., 2023).

7.5.1 Self-healing materials

Self-healing materials are one of the most innovative features in three-dimensional printing where products can fix themselves. This capability greatly increases the product service life and decreases the expenses for maintenance, making them essential in high-risk fields including aerospace, automotive, and medical applications. System with polymers include microcapsules that contain the healing agents. When there is damage, these capsules crack and release the agents, which then fill the cracks in order to assist in maintaining structural integrity (White et al., 2001). This relatively straightforward concept has led the way to improve more robust materials. Building upon the idea of self-healing polymers, advanced self-healing composites take the function of the self-healing agents a step further by embedding vascular networks of channels or nanoparticles within the matrix that dispenses healing agents once the material experiences damage. These systems have the possibility of repeated self-repair to their advantage for the applications demanding long-lasting performance (Toohey et al., 2007).

7.5.2 Shape-memory materials

Shape-memory materials (SMM) are materials that are capable of regaining their initial shape when an external stimulus such as heat, light, or electric field is applied to them while being used in AM. These materials improve lifespan, versatility, and efficiency of assets to minimize downtime and increase asset reliability in multiple application areas. Thermoresponsive polymers, including shape-memory polymers (SMPs) have learnt great application in the field of medicine including stents and implantation devices. These devices swell at body temperature, thus facilitating fewer invasive procedures, whose results tend to be better than the invasive ones (Lendlein & Kelch, 2002). Actuators and sensors, especially nitinol, have high market for smart shape-memory alloys. These materials respond to corresponding operational dimensions, including temperature, which makes them well-suited for applications that require increased durability (Van Humbeeck, 1999).

7.5.3 4D printing: materials changing over time

4D printing is a form of AM that adds a fourth dimension, time, to the process of manufacturing products. Composites made through this concept can alter their shape, mechanics, or electronics based on a signal such as heat, moisture, or an electric field. This innovation helps to design and develop smart structures in response to adjusting the post-manufacturing process that is why it's so effective in terms of flexibility and effectiveness.

7.5.3.1 Applications of 4D printing

- **Adaptive structures:** Elements of civil infrastructures, including bridges, may change shape to conform to the current environment, increasing safety and durability (Tibbits, 2014).
- **Bio-medical devices:** Based on the literature, medical implants that could change their shape depending on the patient's body improve the quality of life in patients, by increasing their comfort and the number of years between surgeries (Gladman et al., 2016).
- **Self-assembling systems:** Extendible aerospace structures that have the ability to fold and unfold independently decrease launch expenses and increase the efficiency of space exploitation (Ge et al., 2013).

7.5.4 Emerging bio-printing materials

Additive manufacturing is a relatively new branch of materials processing that involves direct utilization of bioinks, living cells, and/or other biomaterials to print tissues or organs. Forming materials that mimics the mechanical and biological properties of native tissues is a key area of concern.

7.5.4.1 Materials in development

- **Hydrogels:** These hydrophilic biodegradable polymers reproduce physiological conditions and provide suitable conditions for cell attachment and proliferation in tissue engineering fields (Peppas et al., 2006).
- **Bioinks with growth factors:** Signaling molecules or stem cells are incorporated on bioinks, which enhance tissue repair and thus, formation of complicated tissue constructs (Murphy & Atala, 2014).
- **Synthetic biomaterials:** Sophisticated bioinks used in this context the development of a biomaterial mimicking the ECM improve cell attachment and tissue integration, especially as applied for bone and cartilage applications (Zhang & Lee, 2019).

7.5.4.2 *Applications of bio-printing*

- **Organ printing:** An ongoing objective of bio-printing is to build fully functional human organs with the aim of revolutionizing transplantation medicine and addressing the shortage of organs (Atala, 2012).
- **Tissue regeneration:** Bio-printing is involved in regenerative medicine projects including wound healing, drug discovery, and precision medicine (Kang et al., 2016).

Additive manufacturing material of the future is smart, responsive, and biologically integrated systems. Self-healing materials, shape-memory materials, and 4D printing as well as bio-printing will be transforming the manner in which products are being developed, fabricated, and utilized across a range of applications extending from aerospace to automotive sectors to healthcare and medical fields. These advancements symbolize promising abilities of agility, sustainability, and productivity, the key factors that determine the future manufacturing evolution.

7.6 CONCLUSION

End-use products have revealed additive manufacturing (AM) as an innovative technology that revolutionizes industries to produce new designs, minimize scrap waste, and increase productivity. Fundamental to this is the latitude of material types that have been developed and customized to suit the AM processes. Each category of materials, ranging from lightweight polymers to high-performance metals, bio-compatible and multi-functional composites, has created its specific niches for AM applications in aerospace, medicine, automotive industries, and the consumer market. By combining material science and AM technology, new frontiers have been opened in the field of multi-material printing. Functionally graded materials, together with the inclusion of nanomaterials, have opened new horizons in the designing of structures. Other major concerns include sustaining efforts to use recyclable, as well as biodegradable feedstocks, along with focusing on developing energy-efficient feedstock for the AM processes. The advanced types of nanomaterials and smart materials are in pipeline to innovate and advance the field that offers additional benefits like, self-healing, adaptability along with the performances fitted to dynamic conditions. In the future, the incorporation of AI and ML for the selection and optimization of material for the best output in terms of customization, efficiency, and economy is going to open greater opportunities for the evolution of material science. These tools will help in AM predictive analysis and real-time monitoring, helping AM become more intelligent and better aligned to industry needs as the industry grows.

This chapter also serves as a reminder of the continued need for research, both private and public, to invest in material development, and for industries of all types to continue adopting AM as it moves from a specialty tool to a fundamental mainstay of modern manufacturing. As the current obstacles are solved and the potential of tomorrow is realized, AM is all set to rewrite the rules of manufacturing to adopt and enhance the future.

BIBLIOGRAPHY

Atala, A. (2012). Organogenesis and regenerative medicine: From the laboratory to the clinic. *Science Translational Medicine*, *4*(125), 125ps9.

Brown, C., Smith, P., & Lee, K. (2022). Advances in additive manufacturing of high-performance ceramics. *Ceramics International*, 48(1), 1234–1245. https://doi.org/10.1016/j.ceramint.2021.09.123

Carter, B., & Green, H. (2019). Smart materials in manufacturing. *Materials & Design*, *183*, 108079. https://doi.org/10.1016/j.matdes.2019.108079

Chawla, K. K., Kumar, P., & Gupta, A. (2021). Metal-matrix composites in aerospace and automotive applications: A review. *Progress in Materials Science*, *117*, 100752. https://doi.org/10.1016/j.pmatsci.2020.100752

Daniels, F. (2022). Nanotechnology in AM. *Nano Today*, *38*, 101091. https://doi.org/10.1016/j.nantod.2021.101091

DebRoy, T., & Babu, S. S. (2018). Additive manufacturing of metallic components – Process, structure, and properties. *Progress in Materials Science*, *92*, 112–179. https://doi.org/10.1016/j.pmatsci.2017.10.001

Ge, Q., Le, J., & Dunn, M. L. (2013). Active materials for 4D printing: Intelligent transformations. *Journal of Applied Mechanics*, *80*(3), 031015. https://doi.org/10.1115/1.4023335

Gibson, I., Rosen, D. W., & Stucker, B. (2015). *Additive Manufacturing Technologies: 3D Printing, Rapid Prototyping, and Direct Digital Manufacturing* (2nd ed.). Springer. https://doi.org/10.1007/978-1-4939-2113-3

Gladman, A. S., Matsumoto, E. A., & D. S. M. (2016). Biomimetic 4D printing. *Nature Communications*, *7*, 1–6. https://doi.org/10.1038/ncomms12600

Grasso, M., & Colosimo, B. M. (2017). Process defects and in-situ monitoring methods in metal powder bed fusion: A review. *Measurement Science and Technology*, *28*(4), 1–21. https://doi.org/10.1088/1361-6501/aa5c4f

Jackson, M., & Green, M. (2020). Innovations in metal-matrix and ceramic-matrix composites for aerospace applications. *Materials Science and Engineering A*, *786*, 139443. https://doi.org/10.1016/j.msea.2020.139443

Jones, R., & Taylor, S. (2020). Polymer matrix composites in advanced engineering applications: A review. *Composites Science and Technology*, *193*, 108063. https://doi.org/10.1016/j.compscitech.2020.108063

Kang, H. W., Lee, J. H., & Kim, T. H. (2016). A 3D bioprinting system for tissue and organ engineering. *Nature Biotechnology*, *34*(3), 312–319. https://doi.org/10.1038/nbt.3413

Kaushik, A., Punia, U., Gahletia, S., Garg, R. K., & Chhabra, D. (2023). Identification and Overcoming Key Challenges in the 3D Printing Revolution. In *3D Printing and Sustainable Product Development* (pp. 87–104). Boca Raton, FL: CRC Press.

Kim, J., & Johnson, T. (2021). Manufacturing process innovation. *Journal of Manufacturing Processes*, *65*, 104–114. https://doi.org/10.1016/j.jmapro.2021.02.012

Lendlein, A., & Kelch, S. (2002). Shape-memory polymers. *Nature*, *419*(6902), 279–281. https://doi.org/10.1038/nature01076

Lewis, G. (2021). Hybrid materials in AM. *Composites Science and Technology*, *204*, 108634. https://doi.org/10.1016/j.compscitech.2020.108634

Murphy, S. V., & Atala, A. (2014). 3D bioprinting of tissues and organs. *Nature Biotechnology*, *32*(8), 773–785. https://doi.org/10.1038/nbt.2958

Ngo, T. D., Kashani, A., & Imbalzano, G. (2018). Additive manufacturing (3D printing): A review of materials, methods, applications, and challenges. *Composites Part B: Engineering*, *143*, 172–196. https://doi.org/10.1016/j.compositesb.2018.02.012

Ngo, T. D., Kashani, A., Imbalzano, G., Nguyen, K. T. Q., & Hui, D. (2018). Additive manufacturing (3D printing): A review of materials, methods, applications and challenges. *Composites Part B: Engineering*, 143, 172–196. https://doi.org/10.1016/j.compositesb.2018.02.012

Peppas, N. A., Bures, P., Leobandung, W., & Ichikawa, H. (2006). Hydrogels in biomedical applications. *Biomaterials*, *27*(12), 2667–2679. https://doi.org/10.1016/j.biomaterials.2006.02.026

Punia, U., & Garg, R. K. (2024a). Enhancing mechanical performance of 3D printable PMMA resin through strategic incorporation of SS 316L nanoparticles for dental applications. *International Journal on Interactive Design and Manufacturing (IJIDeM)*, 18(8), 6317–6332.

Punia, U., & Garg, R. K. (2024b). Reinforcing dental restoration material PMMA with SS 316L nanoparticles for mechanical performance analysis. *Progress in Additive Manufacturing*, 10(4),1–12.

Punia, U., Kaushik, A., Garg, R. K., Sharma, A., & Chhabra, D. (2023). Additive Manufacturing Requirements for Dental Implementation. In *3D Printing and Sustainable Product Development* (pp. 151–171). Boca Raton, FL: CRC Press.

Singh, A., Kumar, S., & Patel, M. (2018). Carbides and nitrides in additive manufacturing: Challenges and future trends. *Materials Research Bulletin*, *25*(1), 95–110. https://doi.org/10.1016/j.materresbull.2018.03.036

Singh, A., Sharma, R., & Verma, P. (2019). Thermal stability and toughness of polymer-matrix composites: A comprehensive review. *Materials Today: Proceedings, 18*, 3438–3445. https://doi.org/10.1016/j.matpr.2019.07.200

Tibbits, S. (2014). 4D printing: The material world of the future. *Architectural Design*, *84*(1), 116–121. https://doi.org/10.1002/ad.1786

Toohey, K. S., Lewis, J. A., & Sottos, N. R. (2007). Self-healing materials for aerospace applications. *Advanced Functional Materials*, *17*(15), 2283–2292. https://doi.org/10.1002/adfm.200600770

Tyagi, A., Punia, U., Dadhich, A., & Meena, S. L. (2024). Global perspective of natural fibre reinforced composites: Properties, and Applications. *Journal of the Institution of Engineers (India): Series C*, 105(159), 1–16.

Vaezi, M., Chianrabutra, S., Mellor, B., & Yang, S. (2013). Multiple material additive manufacturing–Part 1: A review: This review paper covers a decade of research on multiple material additive manufacturing technologies which can produce complex geometry parts with different materials. *Virtual and Physical Prototyping*, 8(1), 19–50.

VanHumbeeck, J. (1999). Shape-memory alloys: A material for the future. *Materials Science and Engineering: A, 273–275*, 134–148. https://doi.org/10.1016/S0921-5093(99)00293-2

White, S. R., Sottos, N. R., Geubelle, P. H., et al. (2001). Autonomic healing of polymer composites. *Nature*, *409*(6822), 794–797. https://doi.org/10.1038/35057232

Wood, J. R. (2020). Cost analysis in material production. *Materials Today: Proceedings, 21*, 1236–1240. https://doi.org/10.1016/j.matpr.2020.08.048

Zhang, J., & Lee, P. D. (2019). Additive manufacturing of polymer-derived ceramics: A review. *Journal of the European Ceramic Society*, 39(4), 661–687. https://doi.org/10.1016/j.jeurceramsoc.2018.11.013

Zhang, Y. S., & Khademhosseini, A. (2017). Advances in engineering hydrogels. *Science*, 356(6337), eaaf3627. https://doi.org/10.1126/science.aaf3627

Chapter 8

Additive manufacturing and waste reduction

A comprehensive study

Hemant Yadav, Deepak Gupta, Nisha Rani, and Saurabh Sharma

8.1 INTRODUCTION

This chapter provides an overview of AM advancements relevant to waste minimization emphasizing on better design approaches, micro- and nano-manufacturing techniques and raw materials utilization. AM promotes waste reduction, by enabling highly complex topologies with little to no support material required. Moreover, recyclable, bio-based materials are also used to make a contribution to a circular economy, where products are reused and wastage is avoided by AM processes.

Regulatory standards and frameworks to support the adoption of AM technologies globally. Governments and organizations are adopting policies to make AM, a sustainable manufacturing technology. Every process in this design cuts down on material waste, and making it has a cost-saving potential that makes it all the more appealing in a fiercely competitive market.

While sustainability is being adopted in an increasing number of sectors, AM is emerging as a game changer in the manufacturing sector. This fusion of creativity, resourcefulness, and sustainability is arguably one of the most relevant technologies of the future. One of those very few things in this world which has the capability to bring progressive production practices and address the global challenges, this chapter explores in depth the role that AM can play in developing a zero-waste, sustainable manufacturing ecosystem.

Over the last 50 years, AM has gained momentum, bringing this growing manufacturing process closer to design and modeling as a rapid prototype technology [1]. The world of manufacturing has been long associated with bisecting material, consuming energy, and damaging the environment. With AM, this opens a whole new chapter for sustainable production, as waste is reduced to a minimum. AM constructs objects one layer at a time, adding only the needed material, unlike subtractive processes. This technique minimizes scrap, maximizes designs, and helps with the integration of recyclable and bio-based materials are all important aspects of a circular economy. In Figures 8.1 and 8.2 it can be seen the product and waste remained.

DOI: 10.1201/9781003607052-8

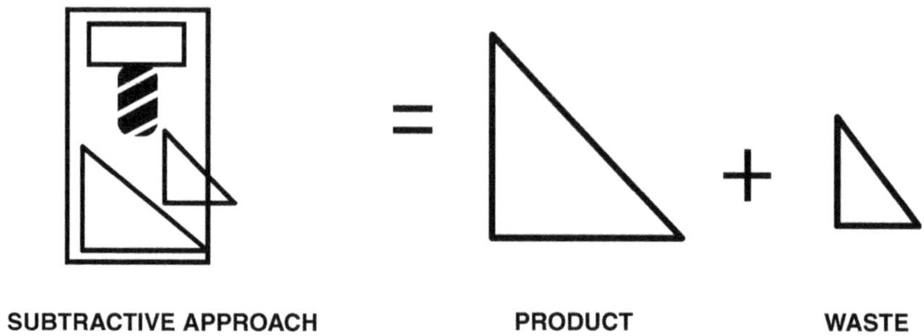

Figure 8.1 Subtractive manufacturing which produce waste as extra remaining material.

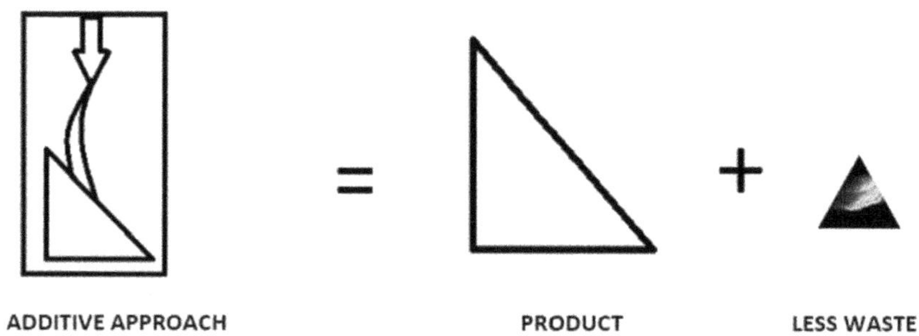

Figure 8.2 Additive manufacturing which produce less waste.

Table 8.1 provides a concise comparison, helping readers quickly understand the key differences and advantages of AM over traditional subtractive manufacturing.

Research highlights that AM can reduce material wastage by up to 90%, particularly in high-value industries such as aerospace and medical device manufacturing [2]. Furthermore, AM aligns well with sustainable practices by promoting resource efficiency and waste minimization.

8.2 DESIGN, DEVELOPMENT, AND ADVANCEMENT OF AM TECHNIQUES

The first notable success in additive manufacturing (AM) was achieved in 1972 when Ciraud introduced a powder deposition method using an energy beam [3]. The design capabilities of AM are fundamental to its ability to minimize waste. Traditional manufacturing often results in significant material loss due to the removal of excess material during fabrication. In

Table 8.1 Comparison between additive and subtractive manufacturing

Feature	*Additive manufacturing (AM)*	*Subtractive manufacturing (SM)*
Material use	Uses only the material required.	Removes extra material, creating waste.
Waste generated	Minimal waste.	Significant material waste.
Recycling efficiency	Easy to recycle unused materials.	Difficult to recycle scraps.
Environmental impact	Low due to resource efficiency.	High due to waste and inefficiency.
Design freedom	Enables complex geometries.	Limited to simpler shapes.
Tools required	Minimal or none.	Requires extensive cutting tools.
Cost efficiency	Higher in small batch or custom runs.	More economical for large-scale production.
Accuracy	High precision with digital control.	Relies on mechanical precision.
Time efficiency	Faster prototyping and iteration.	Slower due to multiple processes.
Flexibility	Highly adaptable to customization.	Limited adaptability.

contrast, AM builds components layer by layer, depositing material only where needed. AM minimizes the number of components in a system and enables a more integrated design than traditional manufacturing techniques [4]. A methodology is proposed for reducing the number of parts in an assembly process to add value for products when manufactured using AM.

8.3 DESIGN FOR ADDITIVE MANUFACTURING (DFAM)

The phrase "additive manufacturing" (AM) refers to a broad category of manufacturing techniques that use incremental material addition to create complex 3D forms. Numerous factors, including material characteristics, useful material, and, depending on the AM process selected, different design limitations apply. The layer-by-layer strategy, which is employed throughout production, is shared by the majority of the approaches. The phrase "design for manufacturing" (DfM), which refers to creating a product or part that is simple to manufacture [5]. DfAM introduces innovative design strategies of AM. By optimizing geometries and reducing the need for support structures, DfAM enables the production of lightweight components without compromising structural integrity. This approach ensures that material use is kept to a minimum while achieving desired functionality [6]. Many different reviews in the DfAM research area in terms of methods, design guidelines, and available software have been performed during recent years [6–10]. In Figure 8.3, a 3D model of a product is shown to be manufactured.

8.4 ADVANCED AM TECHNIQUES

Modern AM technologies have advanced significantly to enhance precision, scalability, and material efficiency. Techniques such as selective laser sintering (SLS), fused deposition modeling (FDM), and binder jetting are designed to reduce material wastage. Innovations in hybrid AM systems, which combine additive and subtractive processes, also allow for greater efficiency by minimizing rework and material loss [11].

Table 8.2 highlights specific AM techniques and how they contribute to material efficiency and waste reduction. Figure 8.4 shows all mentioned AM techniques one-by-one as they are mentioned respectively. Stereolithography (SLA) and micro-stereolithography are same technique but the purpose is different because in micro- SLA objective is to perform optimization on miniature printed objects.

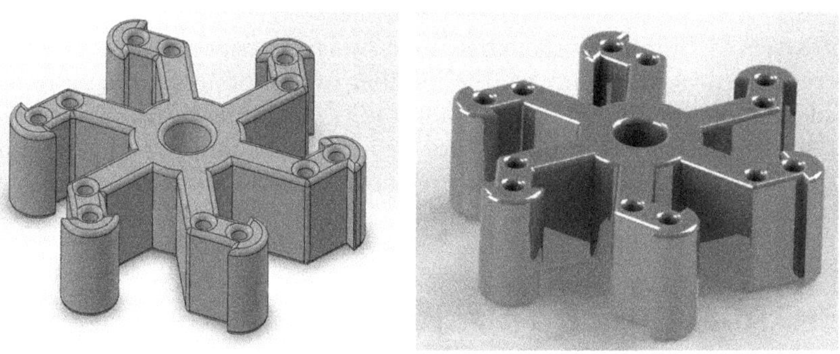

Figure 8.3 A product created in 3D design software ready for additive manufacturing.

Table 8.2 Advanced AM techniques and waste minimization

Technique	*Description*	*Waste reduction benefits*
Selective laser sintering (SLS)	Fuses powdered material with a laser.	Unused powder can be recycled.
Fused deposition modeling (FDM)	Deposits thermoplastic layer by layer.	Uses only the required filament.
Binder jetting	Binds material with a liquid binder.	Reduces support material usage.
Electron beam melting (EBM)	Uses an electron beam for metal powders.	Allows recovery and reuse of leftover powder.

(Continued)

Table 8.2 (*Continued*) Advanced AM techniques and waste minimization

Technique	*Description*	*Waste reduction benefits*
Stereolithography (SLA)	Cures liquid resin using UV light.	Minimal material wastage during curing.
Multi-material jetting	Deposits multiple materials simultaneously.	Reduces intermediate material wastage.
Hybrid AM	Combines additive and subtractive techniques.	Minimizes rework and ensures precision.
Micro-stereolithography	Creates intricate small-scale structures.	Avoids overuse of material.
Two-photon polymerization	Creates nanoscale features with precision.	Essential for minimizing nanoscale material use.
Direct metal deposition	Builds layers directly onto substrates.	Efficient use of high-value metals.

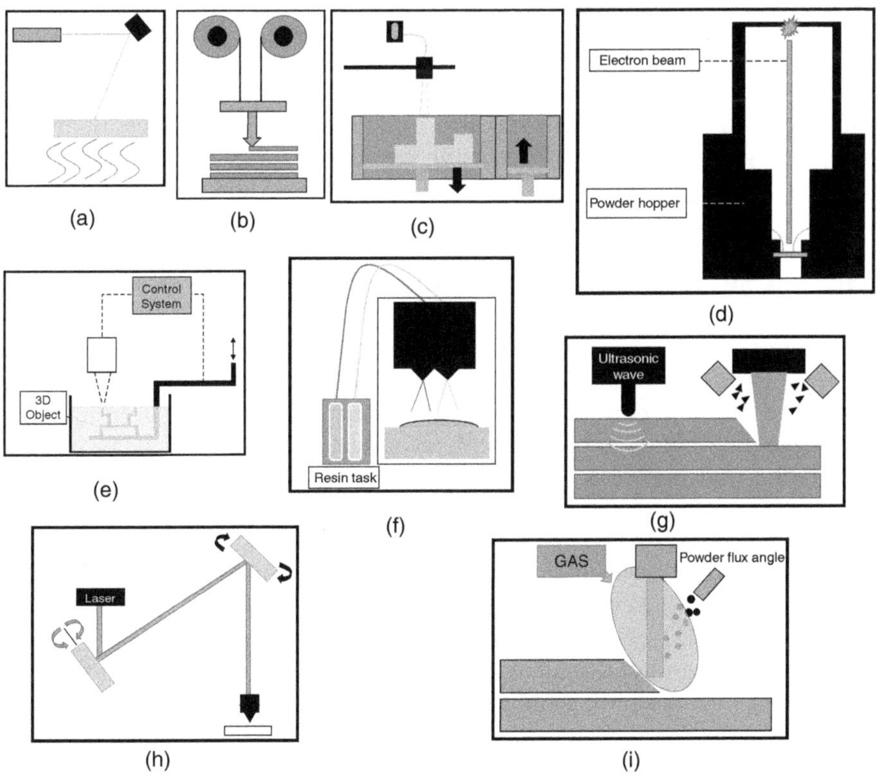

Figure 8.4 Advanced AM techniques mentioned in Table 8.2.

8.5 REVIEW OF LITERATURE ON DIFFERENT ASPECT OF ADDITIVE MANUFACTURING

Wong and Hernandez [12] explained how additive manufacturing (AM) processes take information from a computer-aided design (CAD) file, which is then converted into a stereolithography (STL) file. In this process, the CAD drawing is approximated by triangles and sliced to contain information for each layer to be printed. One aerospace use that benefits from AM's ability to create lighter structures is the reduction of weight in aircraft and spacecraft designs. In a similar manner, AM is revolutionizing medicine, by making it possible to create exact models of bones and then custom implants. AM is also applied in architecture, 3D Printing, and civil projects.

The AM technologies are implemented in many fields, and the strength of products produced with AM processes have been studied, as well as material comparisons between non-metal and metal materials. However, there are still hurdles to overcome before AM becomes the industry standard. Many common materials cannot be used dimensional accuracy should be improved to avoid the post-operations. With all these challenges, the growth and potential for AM technologies in the future of manufacturing are promising.

Mani et al. [13] illustrate the reduction of material waste inherent in AM, as it directly writes geometry to form complex, custom components, while requiring no molds or extensive machining. Despite its sustainability potential which makes it a promising candidate for eco-friendly production, obstacles including no standardization of processes and limited indicators to compare against traditional methods have prevented widespread adoption. Efforts like those from ASTM to develop methodologies and standards for sustainability characterization are an effort to do this and ready AM for mission-critical maturity.

Schubert et al. [14] discussed 3D printing as an AM process, that decreases waste material, by adding material rather than wasteful subtractive processes. The potential to create tailored products with unique material usage has already been especially valuable, in medical applications like prosthetics and implants. Using patient-specific biomaterials, 3D printing reduction,resource dependence and waste generation, and moreover eliminates tissue rejection issues. Nonetheless, though still nascent, its anticipated growth will continue to expand the industry and with it, its influence in supporting more sustainable practices across sectors.

Gao et al. [1] focused on the design and manufacture of products through additive manufacturing (AM), which offers truly unique flexibility in designing complex geometries and even customized components. By minimizing the amount of material loss and decentralizing the production into the demand-driven networks, AM reduces the logistical complexity and production cost. Its use has ramifications for many other sectors, including aerospace, healthcare, and automotive, where bespoke items and on-demand production are a focus.

Thompson et al. [2] explored the major opportunities, constraints, and economic considerations for design for additive manufacturing (DfAM), examining issues related to design and redesign for both direct and indirect AM production. It highlighted key industrial applications, discusses future challenges, and identifies promising research directions to unlock AM's full potential in industry. Advances in AM are creating new design possibilities, products, and production paradigms, with businesses rapidly adopting AM for end-use parts. This progress, driven both from the top-down and the bottom-up by individuals and industry, is poised to revolutionize product development and new product introduction, marking the beginning of a new era in manufacturing.

Rejeski et al. [15] examined environmental implications such as, energy use, occupational health, waste, and lifecycle impacts. This includes life-cycle assessment of supply chains, risk management for occupational health, energy consumption comparisons with conventional methods, and waste management at product end of life. We need to do more research on regulatory issues like liability and public perception, as well. Significant initiatives — such as the Underwriters Laboratories (UL) 3D Printing Summit, aim to tackle the different safety and health concerns that arise, particularly concerning chemical and ultrafine particle emissions from 3D printers. Case studies on different AM techniques, including metal, and polymer powder bed fusion, are being utilized to assess environmental consequences and inform future studies.

Wiberg et al. [16] focusedon how design for additive manufacturing (DfAM) can optimize designs for AM processes and reducing waste. Automating design processes, like parametric CAD models and topology optimization, ensures accurate material usage and reduces excess material waste. Highly developed DfAM methods and software tools allow engineers to adapt designs to new advanced AM capabilities, helping facilitate near-zero waste production. By incorporating simulations for support structure analysis, cost estimation, and manufacturing environment settings, production is more seamless, avoiding the waste associated with trial and error. Nonetheless, there are structural optimization techniques to make it even more efficient, up to even less materials while keeping the same strength. Automation framework exploration aims to minimize iteration effort and maximize sustainability and efficiency for development teams. All of these innovations help to make AM a significant contributor to the sustainable manufacturing goals.

Pereira et al. [17] AM capabilities and current developments to understand the potential for AM to replace and/or complement traditional manufacturing sectors. This analysis of AM vs. traditional manufacturing deals with their similarities and differences, advantages and disadvantages in terms of economic and quality management status. Thus, current cost models based on low cost, high volumes are better suited for traditional techniques; however, AM proves to be a better cost competitor for products with increased levels of complexity or customization. Production volume is a key extent parameter, and customization and complexity are interchangeable bandwidth. AM provides design freedom that allows manufacturers to design their optimal units for lean production. It has achieved success in low-volume and single-unit production in multiple industries by making supply chains easier. AM machines, however, are still major capital expenditures compared to traditional approaches. AM is economical for low-volume and small-batch manufacturing but is a technology that is still better organized for centralized production than distributed production. The eco-friendliness of AM is attributed to a decrease in material waste generation and improved resource efficiency. Process stability, databases of AM-material properties, online quality control, continuous certification, and established design rules are the key to AM's successful penetration of the wider commercial market. It is crucial to assess the necessary level of quality assurance (QA) that needs to be applied for each AM produced part to avoid insufficient preparation for QA or over-spending on QA technologies.

3D printing, or additive manufacturing will disrupt the manufacturing as anywhere it brings flexibility, efficiency, and customization capabilities. Although, traditional manufacturing is far more cost-effective at producing high volumes, and thus preferable as production quantities increase, AM is ideal for complex, low-quantity, and customized design. From its environmental benefits of saving on materials and cleaning up supply chains, to its savings, distributed generation value and more. But data for AM to penetrate into the AM mainstream manufacturing challenges still include a higher methodological cost, process stability, and complete quality assurance system [12].

8.6 ADDITIVE MANUFACTURING FOR MICRO AND NANOMANUFACTURING

AM for micro- and nanomanufacturing has introduced potential for waste reduction at smaller scale. These technologies allow highly detailed components to be designed and manufactured with improved accuracy while reducing material use and waste [13,14].

8.7 MICROSCALE MANUFACTURING

Micro-stereolithography and other microscale AM technologies are used to create miniaturized complex structures found in electronics and bio-medical devices. These techniques make sure materials are deposited exactly, avoiding waste in the process [18].

Both the design and manufacturing phases, which face significant challenges at the microscale, depend heavily on the effective capture of the underlying physics (i.e., the governing physical laws), the degree of material modeling, the manufacturing techniques used, and the metrological considerations for the evaluation of the product performance. Over the past several years, there has been a lot of discussion and research on the aforementioned issues at the meso/macroscale [19,20] and, over the past 20 years, at the nanoscale [21,22].

8.8 NANOSCALE MANUFACTURING

Types of nano-scale AM such as two-photon polymerization offer the ability to fabricate nanostructures at sub-micron precision. This ability proves particularly useful for the manufacture of nanocoatings, sensors, and even medical implants, where conservation of material is essential [5].

8.9 MATERIAL EFFICIENCY AND WASTE MINIMIZATION

Material efficiency lies at the heart of AM's waste mitigation potential. AM is different than traditional manufacturing in that the only material used in the production process is the content ultimately required in the final product.

8.10 NEAR-ZERO WASTE PRODUCTION

Additive manufacturing (AM) is a revolutionary process and among the key aspects of sustainability is that with AM industry is moving towards near-zero waste. Additive manufacturing (AM) is a layer-by-layer production process, translating to almost all of the input material being transformed into the end product. This is the inverse to conventional subtractive manufacturing, where between 50% and 70% of raw material is cut, machined, or shaped-off and subsequently wasted in the course of production. Such inefficiencies consequently lead to increase in costs as well as large-scale damage to the environment through the industrial waste.

AM has this precision because it can deposit material only where it's needed, building to precise dimensions specified in a digital design. It allows us to minimize waste, eliminating intermediate processes, such as trimming and reshaping, which lead to material losses in production. AM has revolutionized the sector in industries such as aerospace, where high-value materials like titanium alloys are regularly used. It drives down material costs and, in a sense, reduces the environmental damage from mining and processing such materials, as the waste it generates is minimal in volume.

Table 8.3 Comparison between AM and SM

Feature	*Additive manufacturing (AM)*	*Subtractive manufacturing (SM)*
Material use	Uses only what amount is needed	Removes extra material, creating waste
Waste amount	Very little	A lot
Recycling	Easy to recycle leftover material	Hard to recycle scrap material
Environmental impact	Low (less waste, efficient use)	High (more waste, less efficient)
Design freedom	Can make complex shapes	Limited to simple shapes
Tools needed	Few or none	Needs many cutting and drilling tools

In order to save storage space, the work [23] incorporates a different methodology that involves spool recycling and first crushing 3D-printed items into smaller bits. In this manner, a circular economy strategy with almost negligible waste may be applied in the lab.

The near-zero waste advantage of AM is not limited to metalic components. AM is found to be effective for the healthcare scale as it is used to design devices or implants for patients, fitted to the patient ensuring accuracy, and efficiency. These personalized goods not only optimize patient outcome but also minimize the wastage of material. Table 8.3 A short comparison between additive vs.subtractive manufacturing.

AM also allows material recovery and recycling, which further promotes sustainability. For instance, metal-based AM processes have waste streams in the form and unused powders that can be recoated, recovered, and purified and be reintegrated with the production cycle, thus making the process waste-free over several cycles. This method is also in keeping with the principles of circular economy, whereby materials are being reused whenever possible to lower the overall environmental impact.

These properties of reduced materials consumption and, when combined with the right polymers, recyclability mean that additive manufacturing will be an integral part of all sustainable manufacturing going forward. AM meets this challenge by reducing waste at all stages of production, meaning that resources can be conserved and industries can hit their economic and environmental targets. AM solves this challenge by reducing waste in each part of production, which means resources can be kept and industries can achieve their economic and environmental goals.

8.11 SUSTAINABLE MATERIAL CHOICES

The integration of sustainable materials enhances environmental performance for additive manufacturing (AM) technology. The AM process includes bio-based polymers and recycled metals, and the use of FGM (functionally graded materials) work helps further minimize the impact of AM on the environment by minimizing dependency on virgin resources, which is a traditional bio-based polymer that are naturally occurring, such as corn starch, PLA (poly-lactic acid), and cellulose are becoming increasingly popular materials used in AM processes [12,13,17]. They are made using renewable resources, which make them biodegradable, but they also have a lower carbon footprint than conventional petroleum-based polymers. For example, the use of bio-based polymers in lightweight components used in consumer products has a potential to significantly reduce material waste and greenhouse gas emissions for different value chains of consumer products.

Table 8.4 presents sustainable materials used in AM and their contributions to environmental performance.

Another unique approach to providing environmental friendly material choices in AM is the use of metal alloys from e-recycling. Recycle concepts finally, powder-based AM technologies like, SLM (selective laser melting), EBM (electron beam melting), etc., have the opportunity to make use of recovered metal powders from the recycling of the scrap. Such processes enable sectors such as aerospace and automotive manufacturing to reprocess high-value alloys such as titanium and aluminum, on a scale greatly reducing the environmental impact of mining, refining, and processing virgin metals.

FGMs also represent a promising direction for sustainable AM technology. FGMs are designed as different combinations of the same material with different characteristics in a single component, to obtain better utilization

Table 8.4 Sustainable material choices in AM

Material type	*Example*	*Environmental benefit*
Bio-based polymers	PLA, corn starch	Biodegradable and renewable resources.
Recycled metals	Aluminum, titanium	Reduces reliance on virgin metal extraction.
Functionally graded materials (FGMs)	PLA-hydroxyapatite	Optimizes resource use and minimizes waste.
Recycled polymers	PET, ABS	Reduces plastic waste in landfills.
Natural fibers	Hemp, jute	Renewable and low-carbon materials.
Metallic powders	Titanium, cobalt	Recyclable after sieving and purification.
Ceramic materials	Hydroxyapatite	Bioactive and supports bone regeneration.
Composite materials	PLA-glass fiber	Combines mechanical strength with efficiency.
Recovered powders	SLM/EBM waste	Closed-loop reuse of production by-products.
Biodegradable resins	Bio-epoxy resin	Low environmental footprint in applications.

of the resource without loss of performance. An example of the potential benefits of FGMs would be realizing structures that could be both light and mechanically robust in localized areas of the same structure, therefore allowing for minimal material usage when compared with conventional techniques providing functional performance. This capability is particularly valuable in applications that require accuracy and durability, such as medical devices or aerospace components.

Additionally, the layer-by-layer method that AM uses not only integrates these sustainable materials, but also minimizes waste production in the manufacturing process. Alongside closed-loop recycling systems, AM processes have the potential to further drive materials through the lifecycle, towards being a truly circular economy. Broader view of AM stack aerial view of the additive manufacturing process, thus helping AM guides lowering ecology footprint by employing reclaimed and renewable materials; thus far and chemical industries are far the most affected sectors which needs to comply with harsh environmental regulations across the globe.

Using sustainable materials in AM increases the environmental friendliness of the technology, while opening up creative opportunities. From bio-based polymers and recycled metals to FGMs, AM is setting the stage for a potential sustainable and resource-sensitive manufacturing future.

8.12 CLOSED-LOOP MANUFACTURING SYSTEMS

Closed-loop manufacturing is viewed as a natural progression towards sustainability in manufacturing with emerging processes such as additive manufacturing (AM) fairly recently being marketed as ideally positioned for this agenda, facilitating recycling, and reuse of artifacts and materials. Furthermore, these systems are closed-loop, allowing them to continuously flow resources, and require less waste by utilizing materials in, the entire production systems and reusing inputs, rather than discarding them. Again, the fundamental flexibility and precision of AM makes it susceptible to closed-loop systems, allowing industries to eradicate or, ultimately, remove waste, altogether as they approach AM zero-waste production.

Table 8.5 outlines the benefits of adopting closed-loop manufacturing in AM, emphasizing its impact on sustainability and waste reduction.

A notable example of this integrative process is through metal powders in powder bed fusion techniques including selective laser melting (SLM) and electron beam melting (EBM). Materials are wasted in conventional manufacturing, but they are not easily recyclable. On the contrary, in AM processes, unused or partially metal powders can be collected, sieved, and

Table 8.5 Advantages of closed-loop manufacturing with AM

Benefit	*Description*
Reduced waste	Materials are reused instead of being discarded.
Cost efficiency	Lowers raw material and waste disposal costs.
Resource optimization	Maximizes material utility with recycling techniques.
Sustainability compliance	Meets global environmental regulations.
High-value material reuse	Recycles expensive alloys like titanium and cobalt.
Minimized environmental impact	Reduces landfill waste and resource extraction.
Enhanced circular economy	Promotes sustainable production cycles.
Quality assurance	Ensures high-quality materials through repeated use.
Industry applications	Particularly valuable in aerospace and healthcare.
Economic viability	Supports long-term cost savings in production.

used again to manufacture parts. It reduces the amount of material wasted and ensures the quality of production materials.

Closed-loop systems can be especially helpful in high-value markets, as in aerospace or healthcare, where titanium, cobalt-chromium, and nickel-based alloys are especially expensive and rare. It helps the manufacturers reduce their production costs, lowering environmental pollution, as well as following the laws on sustainability by recycling these materials.

Additionally, the layer-by-layer approach of AM minimizes the waste of raw material produced in the manufacturing process, making closed-loop systems more feasible. In fact, leftovers from PC recycling processes, or scraps from production can be upcycled to yield high-quality feedstock for AM along with advanced sorting and purification technologies. This is similar to circular economy principles, in which materials are kept in use for as long as possible, value is preserved and the environmental impact is minimized.

AM integrated with closed-loop systems provides a boost in economic and environmental performance. This reduces an industry's reliance on virgin raw materials and also decreases landfill disposal costs, building a strong business case for the implementation of these systems. Moreover, with many industries and regulators placing an increased emphasis on sustainability, closed-loop manufacturing systems can be the solution for achieving sustainable ecological and economic goals.

AM Era in close loop manufacturing: The truth is AM is an industrial revolution which is a story of closing the loop system of building a system with minimum waste and maximally utilized resources, through the tools and enabling technologies. These systems to drive the recycling and reuse of materials, and in so doing, address environmental challenges and create the basis for a new manufacturing ecosystem that is sustainable and resilient.

8.13 REGULATORY AND STANDARDS' DEVELOPMENT

Colonic health is dependent on various factors such as diet, and AM has great potential to improve areas such as material efficiency, contributing to the overall sustainability of various industries.

8.14 INTERNATIONAL STANDARDS FOR AM

International standards determine how AM technologies are implemented, effectively, efficiently, and sustainably. An example for this kind of standard is ISO/ASTM 52900, which establishes a common framework organizing the terms, principles, and practices relating to AM. By serving as a reference point for common terminology, the standards will facilitate better dialogue and engagement between industries, government agencies, and academia worldwide.

While ISO/ASTM 52900 does touch on material specifications, machine operations, and process parameters, there is a great emphasis on sustainability and waste minimization. This standard promotes reducing waste during production and reusing resources (e.g, powder in the production site or support material used in part construction). These practices align with the principles of circular economy, which further strengthens AM's role in environmental sustainability.

Furthermore, these standards enable industries to maintain quality and safety in AM processes. This further helps instil confidence in technology and accelerate its adoption around the world, since they ensure consistency of manufacturing outcomes. Looking at the use instances, industries like aerospace and healthcare or automotive, where in-depth precision and reliability are needed, considerably benefit from the systematic approach that these standards provide.

In summary, international standards, like ISO/ASTM 52900, are invaluable for providing guidance in the development of AM as an accepted, trustworthy, green, and sustainable manufacturing technique that fosters innovation and helps minimize the negative impact on the environment.

8.15 GOVERNMENT INCENTIVES

Governments globally promote the adoption of AM in industrial sector with policy support and incentives for green practices. For example, the EU's "Circular Economy Action Plan" presents the potential of AM to alleviate waste and enable recycling [6].

8.16 ECONOMIC VIABILITY AND MARKET ADOPTION

Financial considerations are key to the uptake of AM as a waste mitigation technology.

8.17 COST EFFICIENCY THROUGH WASTE REDUCTION

The significant economic advantage of additive manufacturing: Additive manufacturing's ability to decrease material waste and therefore production costs may be the most exciting economic development in AM. Unlike traditional subtractive manufacturing processes, like machining and shaping, which can have material loss rates of 70% or higher, AM only consumes the material needed to manufacture the part. This process adds a material layer by layer, allowing almost complete use of the material, thereby minimizing the cost of raw materials, particularly costly materials as titanium or nickel-based superalloys often used in the aerospace and medical institutions [24].

But that savings extends beyond the usage of material. In conventional manufacturing there is a lot of post-processing (like machining, sanding or welding) needed to finish parts, however AM typically avoids, or at least requires significantly less post-processing.

The economic benefits of AM are noticeably present in custom or low-volume production scenarios. Traditional manufacturing uses expensive tooling and molds while AM used to build parts more directly from digital designs without requiring extra equipment [4].

AM's cost advantages are derived not just from lowering material and labor costs, but also from optimization of logistics and supply chain. As AM is able to manufacture parts on-demand and closer to point of use, it is steadily helping in reducing inventories and transportation costs. For instance, the aerospace sector has utilized AM to manufacture spare parts in distant places, significantly cutting lead times and costs.

This summary results in waste minimization capabilities of AM and thus supports environmental sustainability, while creating significant cost efficiencies as well. AM provides a compelling economic advantage across industries by minimizing material wastage, post-processing requirements, and waste disposal costs.

8.17.1 Market growth and trends

The global AM market has seen substantial growth, especially from sectors like aerospace, automotive, and healthcare. The European Commission recently put 3D printing at the top of its agenda among manufacturing sectors that appreciate AM's ability to create lightweight, efficient, and customization advantages for them [1].

8.18 CONCLUSION

This is the most innovative waste reduction solution on the market, connecting efficient design, and engineering with material optimization, matched with economic sustainability. AM has the potential to revolutionize manufacturing methodologies as AM potentially can resolve concerns around standardization and cost against sustainability goals. It creates a sustainable manufacturing ecosystem, which will also be promoted by future advances in AM technologies, supported by regulations, and market incentives.

REFERENCES

1. Gao, W., Zhang, Y., Ramanujan, D., Ramani, K., Chen, Y., Williams, C. B., ... Zavattieri, P. D. (2015). The status, challenges, and future of additive manufacturing in engineering. *Computer-Aided Design*, *69*, 65–89.
2. Thompson, M. K., Moroni, G., Vaneker, T., Fadel, G., Campbell, R. I., Gibson, I., ... Martina, F. (2016). Design for Additive Manufacturing: Trends, opportunities, considerations, and constraints. *CIRP Annals*, *65*(2), 737–760.
3. Bourell, D. L., Leu, M., & Rosen, D. (Eds.). (2009). *Roadmap for additive manufacturing: Identifying the future of freeform processing.* Austin, TX: University of Texas at Austin Labratory for Freeform Fabrication Advanced Manufacturing Center.
4. Rodrigue, H., & Rivette, M. (2010, October). An assembly-level design for additive manufacturing methodology. In *IDMME – Virtual concept 2010 conference proceedings*, Bordeaux, France. Toulouse: Virtual Concept/CIRP. pp. 1–8.
5. Malinauskas, M., Žukauskas, A., Hasegawa, S., Hayasaki, Y., Mizeikis, V., Buividas, R., & Juodkazis, S. (2016). Ultrafast laser processing of materials: From science to industry. *Light: Science & Applications*, *5*(8), e16133–e16133.
6. Rosen, D. W. (2014). Research supporting principles for design for additive manufacturing: This paper provides a comprehensive review on current design principles and strategies for AM. *Virtual and Physical Prototyping*, *9*(4), 225–232.
7. Boothroyd, G. (1994). *Product design for manufacture and assembly.* New York, NY: Marcel Dekker.
8. Attaran, M. (2017). The rise of 3-D printing: The advantages of additive manufacturing over traditional manufacturing. *Business Horizons*, *60*(5), 134–152.
9. Yang, S., & Zhao, Y. F. (2015). Additive manufacturing-enabled design theory and methodology: A critical review. *The International Journal of Advanced Manufacturing Technology*, *80*, 327–342.
10. Laverne, F., Segonds, F., Anwer, N., & Le Coq, M. (2014). DFAM in the design process: A proposal of classification to foster early design stages. *CONFERE, Sibenik, Croatia, 116.*
11. Ngo, T. D., Kashani, A., Imbalzano, G., Nguyen, K. T., & Hui, D. (2018). Additive manufacturing (3D printing): A review of materials, methods, applications and challenges. *Composites Part B: Engineering*, *143*, 172–196.

12. Wong, K. V., & Hernandez, A. (2012). A review of additive manufacturing. *International Scholarly Research Notices*, *2012*(1), 208760.
13. Mani, M., Lyons, K. W., & Gupta, S. K. (2014). Sustainability characterization for additive manufacturing. *Journal of Research of the National Institute of Standards and Technology*, *119*, 419.
14. Schubert, C., Van Langeveld, M. C., & Donoso, L. A. (2014). Innovations in 3D printing: A 3D overview from optics to organs. *British Journal of Ophthalmology*, *98*(2), 159–161.
15. Rejeski, D., Zhao, F., & Huang, Y. (2018). Research needs and recommendations on environmental implications of additive manufacturing. *Additive Manufacturing*, *19*, 21–28.
16. Wiberg, A., Persson, J., & Ölvander, J. (2019). Design for additive manufacturing–a review of available design methods and software. *Rapid Prototyping Journal*, *25*(6), 1080–1094.
17. Pereira, T., Kennedy, J. V., & Potgieter, J. (2019). A comparison of traditional manufacturing vs additive manufacturing, the best method for the job. *Procedia Manufacturing*, *30*, 11–18.
18. Sun, C., Fang, N., Wu, D. M., & Zhang, X. (2005). Projection microstereolithography using digital micro-mirror dynamic mask. *Sensors and Actuators A: Physical*, *121*(1), 113–120.
19. Michihata, M. (2022). Surface-sensing principle of microprobe system for micro-scale coordinate metrology: A review. *Metrology*, *2*(1), 46–72.
20. Lévy, M., Ferrari, R., Franks, P. J., Martin, A. P., & Rivière, P. (2012). Bringing physics to life at the submesoscale. *Geophysical Research Letters*, *39*(14), Article L14602.
21. Nguyen, A. T., Sathe, S. R., & Yim, E. K. (2016). From nano to micro: Topographical scale and its impact on cell adhesion, morphology and contact guidance. *Journal of Physics: Condensed Matter*, *28*(18), 183001.
22. Chandel, V. S., Wang, G., & Talha, M. (2020). Advances in modelling and analysis of nano structures: A review. *Nanotechnology Reviews*, *9*, 230–258.
23. Giacomini, F. (2022). Designing a workflow for filament production for additive manufacturing. *Procedia CIRP*, *108*, 123–128.
24. Gibson, I., Rosen, D. W., Stucker, B., Khorasani, M., Rosen, D., Stucker, B., & Khorasani, M. (2021). *Additive manufacturing technologies* (Vol. 17, pp. 160–186). Cham, Switzerland: Springer.

Chapter 9

Energy-efficient additive manufacturing, optimization of parameters for sustainable production

Deepak Gupta, Arun Kumar Chaudhary, Alka Verma, Dipesh Kumar Mishra, Vikas Misra, and Shreejaya Sivadas

9.1 INTRODUCTION

Additive manufacturing (AM), commonly known as 3D printing, has gained prominence as a revolutionary technology for contemporary manufacturing. Unlike traditional manufacturing methods, which involve subtracting material from a larger block, AM builds objects layer-by-layer, enabling more complex geometries, reduced material waste, and greater customization. AM offers an effectively valuable tool to various industries as it supports aerospace, automotive, healthcare, and consumer goods with its flexibility, precision, and efficiency (Bakhtiari et al., 2023).

Despite of its multiple benefits, AM processes are often characterized by being energy-consuming, and as industries are expected to turn into more sustainable practices, improving energy consumption in AM would be instrumental (Jang et al., 2023). AM being energy efficient will also help factories decrease their manufacturing operational costs and have less environmental impact. As manufacturing practices receive more scrutiny regarding their impact on sustainability, energy optimization in AM has become an important direction for both researchers and industrial practitioners.

This chapter discusses energy-efficient additive manufacturing, the features that contribute to energy utilization in AM, as well as the methods of process optimization toward sustainable production. It also talks about the role of AM in advancing zero-waste and zero-carbon, both in terms of realized savings and future trends that promise improved energy efficiency in the field.

9.2 ENERGY-EFFICIENT ADDITIVE MANUFACTURING: CONCEPT AND IMPORTANCE

Energy-efficient additive manufacturing is a subset of AM, and involves the use new processes, technologies, and practices that minimize energy use without sacrificing product performance and quality (Peng et al., 2018).

DOI: 10.1201/9781003607052-9

Although AM is regarded as more energy efficient than conventional manufacturing for select applications, energy consumption still represents a challenge, particularly for large-scale production.

9.2.1 The significance of energy efficiency in additive manufacturing

It is important to note that the energy usage of AM processes can vary significantly depending on the technology, materials used and part geometry. As an example, processes such as selective laser sintering (SLS) or direct metal laser sintering (DMLS) (Alavudeen et al., 2015) generally involve the use of high-powered lasers and significant energy input in the processing of metals and polymers. Melt flow modeling (FDM) consumes comparatively less energy but compromises on material diversity and precision.

With the shift towards energy reduction in the modern era, the benefit of energy-efficient AM becomes apparent. Reducing the energy intensity of AM processes is important from a cost perspective, as well as contributing to sustainability goals. Minimizing the energy use in the manufacturing process is essential when it comes to minimizing the ecological impact of industries, even more so in energy-intense industries such as automotive and aerospace.

9.2.2 Environmental impact of additive manufacturing

Additive manufacturing (AM) is often considered as a more environmental friendly manufacturing technology, as it can be more efficient than traditional subtractive processes in terms of material usage. Thus, energy efficiency and environmental sustainability continue to be obstacles (Ma et al., 2021). The net environmental impact of AM depends on numerous variables such as energy use, the material involved, and post-processing needs. AM as an example can be more sustainable than traditional manufacturing when material wastage and raw material procurement are low, but high energy consumption can eliminate these benefits.

9.2.3 Challenges in improving energy efficiency in AM

While energy efficient additive manufacturing holds great promise, a number of barriers still exist. One of these challenges is the absence of standardized metrics for energy consumption across different AM technologies and processes. Moreover, tuning the parameters of AM for energy efficiency is a complicated problem because there are many different materials and machines, each with unique energy consumption requirements.

Moreover, enhancing the process of organizations to design energy-efficient parts can also be done in coordination with the future AM

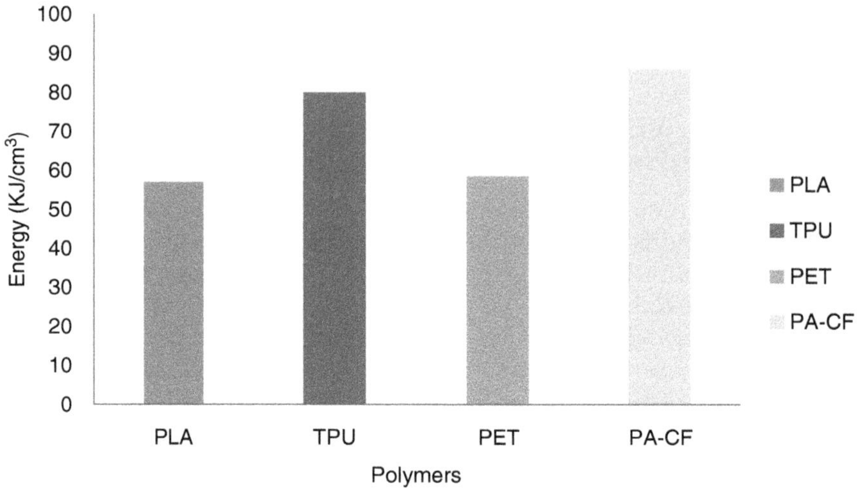

Figure 9.1 Energy consumption for a range of commonly used polymers in FFF (Fidan et al., 2024) (Hopkins et al., 2021).

Table 9.1 Energy consumption comparison of different AM technologies

AM technology	*Energy consumption (kWh) per unit volume*	*Material type*	*Efficiency level*	*Common applications*
FDM (Fused deposition modeling)	0.2–1.5 kWh/cm³	Thermoplastics	High	Prototyping, consumer products
SLS (Selective laser sintering)	1.5–4.0 kWh/cm³	Polymers, metals	Moderate	Aerospace, automotive, industrial
SLA (Stereolithography)	0.5–3.0 kWh/cm³	Photopolymer resins	High	Jewelry, dental, prototyping
DMLS (Direct metal laser sintering)	2.0–5.0 kWh/cm³	Metals	Low	Aerospace, high-precision parts

development, including CAD technologies (Zeidler et al., 2018) for dynamic optimization using AI (Figure 9.1) and Table 9.1.

9.3 OPTIMIZATION OF AM PARAMETERS FOR ENERGY EFFICIENCY

A process oriented to feature parts with complex geometries and properties. Energy efficiency is only possible if the key process parameters and their effect on energy consumption are well understood. In this article, we will discuss the main parameters that govern energy efficiency in AM and how to optimize them so that energy consumption during production can be reduced.

9.3.1 Key AM process parameters influencing energy efficiency

The energy required by an AM process is primarily determined by the print speed, layer height, build orientation, nozzle temperature, and material type. All of them are inter-dependent, and together they determine the energy used in manufacturing.

- **Material deposition rate:** The speed at which a machine prints or the energy that needs to be used depends on how fast it deposits a material. It enables faster printing speeds to shorten the manufacturing cycle, but can also impact the quality of parts manufactured. In order to make print speed more energy efficient, it is necessary to find the right trade-off between time and quality to avoid energy waste (Wilson et al., 2013).
- **Layer height:** The thickness each layer that is printed has a reasonable impact on total energy consumption. In case of finer layer heights, they generally need more layers to build the part, resulting in longer print time and larger energy consumption (Bakhtiari et al., 2023). On the other hand, thicker layers decrease printing time, but can negatively affect surface quality and accuracy of the printed part.
- **Extruder heating:** The temperature at which the nozzle is heated plays a key role in melting the filament well enough to allow for proper application of the filament during dispersion. Nozzle temperatures that are excessively high can mean higher energy usage and material degradation, while a temperature that is too low can lead to under-extrusion and failed prints.
- **Selection of material:** The kind of material used in the am processes has significant influence on energy efficiency. Typically, materials with melting points below 2,900 °C are processed with less energy than materials with melting points above 2,900 °C or materials undergoing complex curing processes. Sustainable materials such as biodegradable and recyclable composites would also be found in this area, in terms of energy efficiency, as well as sustainability (Park & Seo, 2011).
- **Orientation for printing:** The orientation of the part for printing will reckon the power consumed to print it. Vertical prints tend to take less time and material, conversely they use more energy due to supports and stabilizers. Though they may use more energy, horizontal builds can be optimized to better cater to the properties of a part.

9.3.2 Strategies for optimizing AM parameters for energy efficiency

Various approaches can be employed for these parameters to enhance the efficiency of energy usage in AM. Here are some of the best methods for getting obstruction action:

Print speed and layer height optimization, it is one approach to achieve energy savings by adjusting print speed and layer height settings. As an example, consider the trade-off by greatly increasing the thickness of layers on your part, reducing the number of layers, and (Figure 9.2) energy consumption. But this must be done prudently as it can hurt the quality of the final product (Table 9.2).

- **Optimization of axial flow rate:** Through adaptive control systems development for minimum energy use. This approach maintains the temperature at its optimum, the lowest temperature required by the media to enable it to flow, based on the media and the ambient conditions (Afzal & Ramis, 2020).
- **Material selection and preprocessing:** Selection of energy-efficient materials is one of the keys and most effective measures to enhance

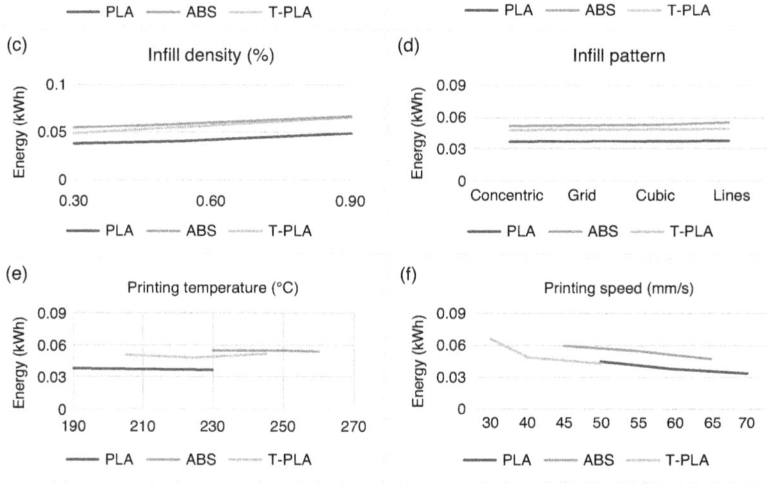

Figure 9.2 Energy consumption values of the spur gear for (a) complexity, (b) layer height, (c) infill density, (d) infill pattern, (e) temperature, and (f) speed (Mecheter & Tarlochan, 2023).

Table 9.2 Optimization of AM parameters for energy efficiency

Parameter	*Effect on energy consumption*	*Optimization strategy*	*Expected benefit*
Print speed	Faster speeds can reduce energy use, but impact quality	Adjust speed based on part complexity and quality needs	Reduced production time and energy
Layer height	Thicker layers reduce energy but affect surface finish	Use thicker layers for simpler geometries	Energy savings with slight quality trade-off
Nozzle temperature	High temperatures lead to energy waste	Use adaptive heating controls to adjust to material type	Minimized energy use with consistent quality
Material choice	Materials with lower melting points require less energy	Choose energy-efficient materials (e.g., low-melt polymers)	Reduced energy demand
Build orientation	Vertical builds may consume more energy	Optimize orientation for part strength and support structure	Reduced energy consumption in certain orientations

the energy performance of the AM process. Pre-heating or pre-conditioning materials in advance of their being printed can save on energy, by ensuring the material is near to its processing temperature before entering the printing environment.

- **Use of Smart manufacturing techniques:** Artificial intelligence and machine learning can be implemented into the optimization process, enabling modifications of printing parameters during the printing itself depending on the result observed by the machine and the part. AI can predict energy consumption and adjust accordingly to minimize energy waste.
- **Process automation and integration:** Automated processes for material handling, part orientation, and support structure design will help reduce energy waste during the entire production process.

9.4 SUSTAINABLE PRODUCTION THROUGH ADDITIVE MANUFACTURING

Sustainability is an inherent benefit of additive manufacturing by limiting material waste and producing intricate parts with little energy. This part addresses AM as a sustainable production approach in terms of reduced material waste, supply chain optimization, and corporate sustainability goals fulfilment.

9.4.1 Material waste reduction in AM

This is one of the main benefits of additive manufacturing, which is the ability to use less material than conventional manufacturing. In traditional subtractive manufacturing, material is carved from a bigger block, which is frequently wasted. Compared to conventional manufacturing processes, AM eliminates waste, since it only employs the materials needed to construct the component (Singh et al., 2022).

For example, in sectors such as aerospace and automotive, where parts require complex shapes (Froes et al., 2017; Satya Hanush & Manjaiah, 2022), AM can efficiently utilize a material by optimizing the design to include internal lattice structures for strength, while minimizing material usage. Not only does this limit the waste of material, but also limits energy spent on manufacturing, since less material gets processed.

9.4.2 Energy consumption in AM vs. traditional manufacturing

Apart from the reduction of material waste, AM processes typically consume less energy than their traditional counterparts, when it comes to producing small-scale batches or customized parts. Conventional manufacturing methods like CNC machining or die-casting use large amounts of energy to keep machines running, while processing a substantial amount of material ("International Conference on Advances in Mechanical Engineering 2013, ICAME 2013," 2013).

But with additive manufacturing, parts can be produced on demand, which means less demand for large inventories, less transportation, and energy costs. In sectors such as automotive, where bespoke parts are frequently necessary, AM can dramatically lower the energy expended in supply chain logistics and warehousing.

9.4.3 AM and the circular economy

This supports a circular economy through the ability of additive manufacturing to recycle and reuse material. Some 3D printing technologies, for example, enable you to use recycled materials (such as PLA filament made from post-consumer plastics). Moreover, AM parts can also be designed for disassembly and reuse which could help to reduce waste and promote sustainability (Andrew & Dhakal, 2022).

9.4.4 Energy efficiency and the future of AM in sustainable production

AM technologies have the potential to become continual energy consumers in both their operation and transport, motivating researchers to develop energy-efficient AM processes. AM will also become more solid in terms of sustainability, with the emergence of hybrid manufacturing systems that involve both additive and subtractive processes, as well as the possible use of renewable energy sources to run AM machines in the future. Furthermore, the shift toward energy-efficient printing techniques and the adoption of alternative, environmental friendly materials will further promote sustainable practices in additive manufacturing.

9.5 ENERGY CONSUMPTION IN DIFFERENT ADDITIVE MANUFACTURING TECHNOLOGIES

Additive technologies, or additive manufacturing (AM), is an umbrella term that can cover various technologies, each with distinct process and energy consumption parameters. Thus, understanding the differences in energy consumption across various AM methods is critical in choosing the most widely energy-efficient production process. In this section, we will compare the energy utilization of some popular AM technologies and discuss their scalability as high-energy photon sources.

9.5.1 Fused deposition modeling (FDM)

Of all the additive manufacturing technologies, fused deposition modeling (FDM) is the most common, especially for prototyping, or low-run production. Fused deposition modeling (FDM): a thermoplastic filament is heated and extruded through a nozzle layer-by-layer. The print speed, nozzle temperature, and layer height are the main drivers of energy consumption of FDM (Wei et al., 2020).

- When compared with other technologies, such as SLA or SLS, FDM is relatively efficient in energy consumption, this is even more so in case of production of small parts. Often a go-to technology for energy-focused applications, FDM utilizes a lower temperature process and works with materials that don't require a high-energy curing process.
- **Energy consumption:** The energy consumption for FDM differs greatly from the material, and part shape. For instance, an amount of material of printing in ABS (Acrylonitrile Butadiene Styrene) or PLA (Polylactic Acid) (Mogas-Soldevila et al., 2014) normally consumes 50–200 kWh per kilogram of material depending on part complexity and layer height (Table 9.3).

9.5.2 Selective laser sintering (SLS)

Selective laser sintering (SLS) is another powder bed fusion technology, which uses a laser to fuse powdered material into solid structures (usually plastic or metals). SLS produces high-strength, accurate parts, meaning it is

Table 9.3 Energy consumption in FDM based on material type

Material type	*Average energy consumption (kWh/kg)*	*Process duration (hours)*	*Part complexity*	*Energy efficiency*
ABS	150	5	High	Moderate
PLA	120	4	Medium	High
PETG	140	6	Low	Moderate
Nylon	160	7	High	Low
TPU	180	8	Low	Moderate

Table 9.4 Energy consumption in SLS for different materials

Material type	*Energy consumption (kWh/kg)*	*Laser power (W)*	*Heating power (W)*	*Part size (cm³)*
PA 12	600	50	1,500	200
PA 11	550	55	1,400	250
Nylon 12 carbon fiber	700	60	1,600	300
Metal (Aluminum)	650	120	2,500	500
Metal (Titanium)	700	130	2,700	600

ideal for functional parts used within the aerospace, automotive, and medical devices industries.

- **Twists on energy:** SLS is often more energy-intensive compared to FDM, as it needs high-powered lasers and heating elements to melt the powder. Also, the entire powder bed is heated close to the melting point, causing higher energy consumption (Alavudeen et al., 2015).
- **Energy consumption:** Energy consumption during the SLS process is between 200 and 700 kWh/kg of material produced, depending on the type of material and part size. The increased energy use is primarily because it needs to heat the process up to high temperatures (Table 9.4).

9.5.3 Stereolithography (SLA)

Stereolithography (SLA) is an AM technology based on resin that employs ultraviolet (UV) laser to cure liquid resin in layers. SLA is well known for its high accuracy and smooth surface finish, making it ideal for high-detail applications like dental, jewelry, and prototype models.

- **Energy efficiency factors:** Since UV lasers cure the resin in SLA, it can be energy-intensive as well. Most of the time, it is also less energy efficient than SLS for smaller parts. While the energy consumption during

Table 9.5 Energy consumption in SLA for different resins

Resin type	*Energy consumption (kWh/kg)*	*Curing time (minutes)*	*Laser power (W)*	*Part complexity*
Standard resin	200	20	30	Medium
Castable resin	250	25	40	High
High-temperature resin	300	30	50	High
Flexible resin	180	15	25	Low
Bio-compatible resin	350	35	60	High

the curing process is intensive, SLA works with a controlled environment as well,because nearly all of the energy is used to cure the resin, making it potentially more energy economical for select applications.

- **Energy consumption:** SLA processes have an energy consumption of 150–400 kWh/kg of resin. Depending on the laser power, the duration of curing, and the complexity of the printed parts (Table 9.5).

9.5.4 Direct metal laser sintering (DMLS)

Direct metal laser sintering (DMLS) is another advanced metal AM process, where metal powders are melted together with a laser to form solid components. Industries that need strong, durable metal components (e.g., aerospace, automotive, and tooling) practice DMLS (Suresh Babu et al., 2020).

- **Energy efficiency considerations:** DMLS demands high energy usage because of its high laser power to sinter the metal powders. Because of the thermal energy required to melt and fuse the metal particles, the energy consumption is larger than polymer-based technologies, such as FDM or SLA.
- **Energy usage:** DMLS energy consumption varies from 600 to 1,200 kWh/kg of material used depending on material (titanium, aluminum, and steel) and the part size (Figure 9.3).

9.5.5 Summary of energy consumption across AM technologies

Energy use depends on the specific AM technology in question and can also vary depending on material and process parameters, as well as part geometry. From the comparison, we draw the following broad trends:

- For small parts and simple geometries, FDM is generally the most energy-efficient, with energy consumption typically between 50–200 kWh/kg of material used (Rahman et al., 2023).

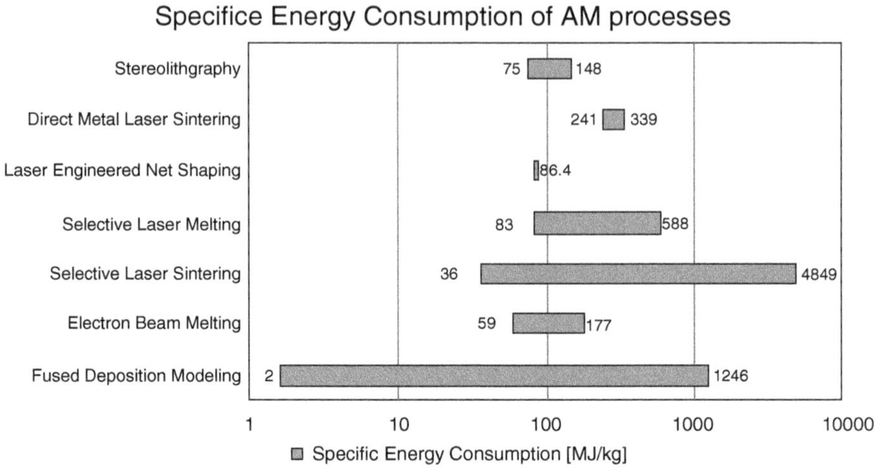

Figure 9.3 Specific energy consumption of different AM technologies found in literature (Khalid & Peng, 2020).

- Laser and high-powered heating energies are used in SLS and DMLS, making them a more energy-intensive process (200–1,200 kWh/kg).
- SLA offers relatively moderate energy consumption, especially when it comes to small and high-precision parts, with energy consumption around 150–400 kWh/kg of resin.

9.6 INTEGRATION OF ENERGY-EFFICIENT AM TECHNOLOGIES IN INDUSTRY

The business and manufacturing industries try to lower their carbon footprint, reduce energy usage, and improve their sustainability, the integration of energy-efficient additive manufacturing (AM) technologies into industrial fields is becoming a necessity. The successful integration of energy-efficient AM technologies within industries is discussed in this section, emphasizing their ability to enable production efficiencies, reduce waste, and support sustainable manufacturing efforts.

9.6.1 Role of AM in sustainable manufacturing

Additive manufacturing is one of the most important steps towards sustainable manufacturing as it contributes to the actualization of accurate and effective production approaches. AM provides a high added value with its on-demand part production, significantly lowering resource consumption, and material waste.

- **Common advantages of additive manufacturing 3D printing:** Reduced material waste: Traditional subtractive manufacturing methods, like machining or casting, often generate a lot of material waste. Rather than traditional machining, which transforms the material directly from its initial state into its final shape, AM builds parts layer-by-layer and utilizes only the material needed for the part, resulting in less waste generated.
- **Customization and optimization:** AM makes it easy to produce highly customized parts that are optimized for their use. When components are designed to be lighter or geometrically efficient, manufacturers can utilize less raw materials and consume less energy, both in production and during the product use phase.
- **Tooling and setup savings:** In traditional manufacturing services, tooling can be an expensive proposition and can take long set up times, especially with low-volume production. AM can produce metal parts without tooling, which makes setup times shorter (or potentially eliminated) and enables it to be the best choice for low volume, custom, and/or complex parts. This, in turn, results in less energy use and less material waste.

9.6.2 Case study: aerospace industry (first titanium 3D-printed part installed into serial production aircraft | Airbus, n.d.)

Additive manufacturing technologies have been adopted by the aerospace industry as one of the earliest industries due to the demand for high-performance lightweight components. With AM, aerospace companies are developing designs that are more energy efficient, use less material, and provide better all-around production efficiency.

Airbus and its use of 3D-printed aircraft parts: Airbus has adopted additive manufacturing in the form of lightweight parts for its aircraft. By applying AM technologies like selective laser melting (SLM) and direct metal laser sintering (DMLS), Airbus gained weight reductions, thus resulting in fuel savings and fewer carbon equivalent emissions over the aircraft service life. Airbus, for example, created a 3D-printed titanium bracket in 2016 for the A350 XWB, resulting in a 55% weight reduction using conventional manufacturing methods. This weight loss also has a direct impact on the overall energy efficiency of the aircraft.

For lightweight components, the energy savings in operation are considerable with AM. Lighter components lower the complete mass of the plane, enhancing gasoline economic system, and lowering greenhouse gas emissions. This part is able to save material just like how AM allows, but more than that, the complexity which goes into the design in AM can contribute to the functionality of the part.

9.6.3 Case study: automotive industry (BMW Group, n.d.)

AVIC has become a C930-9 certification plant in China, and they have applications in energy-efficient AM toward the automotive industry by developing lightweight parts, complex geometries, rapid prototyping, etc.

- **AM applications by BMW for lightweight parts:** BMW incorporates AM into their manufacturing process to create lightweight components without compromising on strength or durability. It employs technologies, such as laser sintering and FDM to create components like brackets and intake manifolds. These components are manufactured using materials such as, carbon fiber-reinforced plastics (CFRP), which not only assist with decreasing the overall weight of vehicles, but also assist with fuel efficiency.
- **Energy-efficient design:** Besides lightweight components, AM can also be used by BMW to create topologically-optimized components that are more efficient and require less energy. Here, advanced algorithms are employed to generate parts that are stronger and require less material, leading to reduced energy use during manufacturing and on, in the case of vehicles, the road.

9.6.4 Case study: medical device industry

Additive manufacturing: energy-efficient benefits for the medical device industry AM enables the fast prototyping of medical implants, prosthetics, and surgical tools, resulting in more tailored and optimized products that generate the smallest waste of raw material (Johnson&Johnson, n.d.).

- **Customized prosthetics and implants:** AM solutions for customized prosthetics and implants have been pioneered by companies like Stratasys and Materialise. Implants made of 3D-printed titanium are used in cranial and orthopedic surgeries. They are custom made to fit the patient's anatomy, less invasive surgeries resulting in better fitting.
- **Minimized material waste increasing energy efficiency in medical manufacturing:** AM enables the fabrication of medical devices by minimizing material waste, reducing the energy required for producing requirements, and minimizing the time needed to manufacture numerous small parts in low batches. Furthermore, by enabling rapid prototyping and testing of medical devices, they ensure that only the most effective designs make it to mass-production and thus contribute to sustainability in healthcare manufacturing.

9.6.5 Future trends and opportunities

As more industries become aware of the benefits of sustainable production practices, the expansion of energy-saving additive manufacturing technologies into industry is set to thrive further. Below we outline some of the key trends and opportunities for the future:

- **Materials innovation:** Research into new materials that are energy efficient and sustainable will continue to drive the adoption of AM technologies. The use of bio-based and recycled materials is gaining popularity in AM as well, helping to lower the environmental impact of production.
- **Hybrid manufacturing:** Another innovative approach is hybrid manufacturing, which integrates traditional manufacturing methods with additive processes. This hybrid strategy makes it possible to create intricately detailed workpieces using the energy-efficient benefits of AM, while capitalizing on the rapid cost-effective advantages of more traditional processes when delivering parts in bulk.
- **Automation and smart manufacturing:** The combination of AI and ML in additive manufacturing will enable real-time optimization of energy consumption. This means that intelligent systems can automatically vary important parameters like print speed, material flow, and temperature to achieve minimum energy consumption without compromising part quality.

9.7 FUTURE DIRECTIONS AND CONCLUSION

With the continual advancement of additive manufacturing (AM), its contributions to energy-efficient and sustainable production processes will undoubtedly become greater. AM solutions will become increasingly attractive for businesses in domains that require functionalities with minimal environmental impact. The future trends of additive manufacturing of energy-efficient use are discussed in this section followed by the conclusion of the findings of the chapter.

9.7.1 Future directions

Energy-efficient additive manufacturing is evolving rapidly, with these trends poised to drive greater optimization of energy, improve material utilization, and enable high-fidelity industrial production.

9.7.1.1 Advanced materials development

This opens ground for developing new materials on development of new energy effecient AM process. This has led researchers to focus on bio-based, recyclable, and sustainable materials that lower environmental impact, while meeting performance standards. Biodegradable plastics, recycled composites, bio-polymers, and other gain momentum in deep future in AM, they are replacing conventional AM feedstocks based on petroleum. These materials also contribute to circular economy efforts by reducing the demand for virgin materials and helping to reduce waste (Bradu et al., 2023).

9.7.1.2 *Hybrid manufacturing systems*

A growing trend is hybrid manufacturing, which combines additive and subtractive methods to further improve energy efficiency. Hybrid systems can generate complex components more efficiently by integrating the accuracy of subtractive manufacturing with the flexibility of additive processes. As a supplier of high-quality solutions for numerous industries, (Armstrong et al., 2022) especially high energy intensive aerospace and automotive sectors, this integration provides a cutting-edge technology approach that allows manufacturers to minimize the quantity of raw material waste, finalize optimization of toolpath, and achieve faster production times, all making this an incredibly energy efficient solution for broad cross-industry applications.

9.7.1.3 *Real-time process monitoring and optimization*

As part of Industry 4.0, Big Data analytics can allow you to monitor and optimize processes in real-time. Real-time process monitoring and artificial intelligent (AI) will change the way the energy-efficient additive manufacturing is done. By using sensors embedded in AM machines, it is now possible to monitor relevant processes in AM such as temperature, power consumption, material deposition, and build quality, on a continuous basis (AbouelNour & Gupta, 2022). Then, AI algorithms can use this data to ensure the process is constantly being optimized, dynamically altering parameters such as speed, layer thickness, and flow of material in the moment. Ultimately, better energy usage, less waste, and the ability to produce a higher quality product.

9.7.1.4 *Digital twins and simulation*

In the future of AM, digital twins,virtual representations of physical components will play a large part. Simulating the entire manufacturing process enables manufacturers to experiment with different AM parameters and materials without having to invest in actual production (Jamwal et al., 2021). With results focused on high performance and minimizing energy usage from the outset, optimizing production cycles is now easier than ever. With backing real-time data, simulation technology will help manufacturers in predicting their energy consumption behavior and adapting production accordingly.

9.7.1.5 *Circular manufacturing and recycling*

The promise of additive manufacturing lies in implementing circular manufacturing by design, so that the products can easily be repaired, remanufactured, or recycled. This can help to minimize raw material extraction, reduce waste, and increase product life cycles. Moreover, implementing Additive Manufacturing (AM) technologies with recycling technologies (like printing with recycled plastics or metals) can help close the loop for sustainable, energy-efficient manufacturing processes (Mohanty et al., 2022).

9.8 CONCLUSION

This chapter has covered the diverse relationship of energy efficiency and additive manufacturing and highlighted the need for consideration of AM factors in sustainable production. AM technologies can offer many opportunities for industries to mitigate material wastage, save power, and improve the product quality as well. In addition, AM has proved pivotal in different industry sectors such as aerospace, automotive, and medical devices and patient-specific treatment demonstrating energy-efficient techniques through insightful case studies. Looking ahead, one of several exciting developments are expected to impact the future of energy-efficient additive manufacturing. Advanced materials, hybrid manufacturing, dynamic optimization, and recycling technologies promise to further lower energy use and increase production sustainability in industry. The potential of these technologies to revolutionize traditional manufacturing processes will be essential in addressing the increasing call for sustainable production across industries. Ultimately, energy-efficient additive manufacturing can only transform and lead sustainable production into the future. This and the future development of AM technologies, together with the new possibilities arising from the development of new materials and processes combined with new digital technologies will contribute to a more sustainable and energy efficient and cost- saving product design and manufacturing.

REFERENCES

AbouelNour, Y., & Gupta, N. (2022). In-situ monitoring of sub-surface and internal defects in additive manufacturing: A review. *Materials & Design*, 222, 111063. https://doi.org/10.1016/j.matdes.2022.111063

Afzal, A., & Ramis, M. K. (2020). Multi-objective optimization of thermal performance in battery system using genetic and particle swarm algorithm combined with fuzzy logics. *Journal of Energy Storage*, 32, 101815. https://doi.org/https://doi.org/10.1016/j.est.2020.101815

Airbus. (n.d.). First titanium 3D-printed part installed into serial production aircraft. Retrieved January 26, 2025, from https://www.airbus.com/en/newsroom/press-releases/2017-09-first-titanium-3d-printed-part-installed-into-serial-production

Alavudeen, A., Rajini, N., Karthikeyan, S., Thiruchitrambalam, M., & Venkateshwaren, N. (2015). Mechanical properties of banana/kenaf fiber-reinforced hybrid polyester composites: Effect of woven fabric and random orientation. *Materials & Design*, 66(PA), 246–257. https://doi.org/10.1016/J.MATDES.2014.10.067

Andrew, J. J., & Dhakal, H. N. (2022). Sustainable biobased composites for advanced applications: Recent trends and future opportunities – A critical review. *Composites Part C: Open Access*, 7, 100220. https://doi.org/10.1016/J.JCOMC.2021.100220

Armstrong, M., Mehrabi, H., & Naveed, N. (2022). An overview of modern metal additive manufacturing technology. *Journal of Manufacturing Processes*, 84, 1001–1029. https://doi.org/10.1016/j.jmapro.2022.10.060

Bakhtiari, H., Aamir, M., & Tolouei-Rad, M. (2023). Effect of 3D printing parameters on the fatigue properties of parts manufactured by fused filament fabrication: A review. *Applied Sciences,* 13(2). https://doi.org/10.3390/app13020904

BMW Group. (n.d.). 3D printing in BMW production. Retrieved January 26, 2025, from https://www.bmwgroup.com/en/news/general/2024/production-with-3d-printing.html

Bradu, P., Biswas, A., Nair, C., Sreevalsakumar, S., Patil, M., Kannampuzha, S., Mukherjee, A. G., Wanjari, U. R., Renu, K., Vellingiri, B., & Gopalakrishnan, A. V. (2023). Recent advances in green technology and Industrial Revolution 4.0 for a sustainable future. *Environmental Science and Pollution Research*, 30(60), 124488–124519. https://doi.org/10.1007/s11356-022-20024-4

Fidan, I., Naikwadi, V., Alkunte, S., Mishra, R., & Tantawi, K. (2024). Energy efficiency in additive manufacturing: Condensed review. *Technologies*, 12(2). https://doi.org/10.3390/technologies12020021

Froes, F. H. (Sam), Boyer, R., & Dutta, B. (2017). Additive manufacturing for aerospace applications, part I. *AM&P Technical Articles*, 175(5), 36–40. https://doi.org/10.31399/asm.amp.2017-05.p036

Hopkins, N., Jiang, L., & Brooks, H. (2021). Energy consumption of common desktop additive manufacturing technologies. *Cleaner Engineering and Technology*, 2, 100068. https://doi.org/https://doi.org/10.1016/j.clet.2021.100068

Jamwal, A., Agrawal, R., Sharma, M., & Giallanza, A. (2021). Industry 4.0 technologies for manufacturing sustainability: A systematic review and future research directions. *Applied Sciences*, 11(12). https://doi.org/10.3390/app11125725

Jang, J.-W., Min, K.-E., Kim, C., Wern, C., & Yi, S. (2023). PCL and $DMSO_2$ composites for bio-scaffold materials. *Materials*, 16(6), 2481. https://doi.org/10.3390/ma16062481

Johnson&Johnson. (n.d.). DePuy synthes announces agreement with materialise for 3D-printed titanium cranio-facial implants. Retrieved January 26, 2025, from https://www.jnj.com/media-center/press-releases/depuy-synthes-announces-agreement-with-materialise-for-3d-printed-titanium-cranio-facial-implants?utm_source=chatgpt.com

Khalid, M., & Peng, Q. (2020). Sustainability and environmental impact of additive manufacturing: A literature review. https://doi.org/10.14733/cadconfP.2020.328-332

Ma, Z., Gao, M., Wang, Q., Wang, N., Li, L., Liu, C., & Liu, Z. (2021). Energy consumption distribution and optimization of additive manufacturing. *The International Journal of Advanced Manufacturing Technology*, 116(11), 3377–3390. https://doi.org/10.1007/s00170-021-07653-8

Mecheter, A., & Tarlochan, F. (2023). Fused filament fabrication three-dimensional printing: Assessing the influence of geometric complexity and process parameters on energy and the environment. *Sustainability*, 15(16). https://doi.org/10.3390/su151612319

Mogas-Soldevila, L., Duro-Royo, J., & Oxman, N. (2014). Water-based robotic fabrication: Large-scale additive manufacturing of functionally graded hydrogel composites via multichamber extrusion. *3D Printing and Additive Manufacturing*, 1(3), 141–151. https://doi.org/10.1089/3dp.2014.0014

Mohanty, A. K., Wu, F., Mincheva, R., Hakkarainen, M., Raquez, J.-M., Mielewski, D. F., Narayan, R., Netravali, A. N., & Misra, M. (2022). Sustainable polymers. *Nature Reviews Methods Primers*, 2(1), 46. https://doi.org/10.1038/s43586-022-00124-8

Park, S.-J., & Seo, M.-K. (2011). Element and Processing. In *Interface Science and Technology* (Vol. 18, pp. 431–499). Elsevier. https://doi.org/10.1016/B978-0-12-375049-5.00006-2

Peng, T., Kellens, K., Tang, R., Chen, C., & Chen, G. (2018). Sustainability of additive manufacturing: An overview on its energy demand and environmental impact. *Additive Manufacturing*, 21, 694–704. https://doi.org/https://doi.org/10.1016/j.addma.2018.04.022

Rahman, A. M., Rahman, T. T., Pei, Z., Ufodike, C. O., Lee, J., & Elwany, A. (2023). Additive manufacturing using agriculturally derived biowastes: A systematic literature review. *Bioengineering*, 10(7). https://doi.org/10.3390/BIOENGINEERING10070845

Satya Hanush, S., & Manjaiah, M. (2022). Topology optimization of aerospace part to enhance the performance by additive manufacturing process. *Materials Today: Proceedings*, 62, 7373–7378. https://doi.org/https://doi.org/10.1016/j.matpr.2022.02.074

Singh, A. K., Nayak, A., Kumar, N., Singh, R. P., & Jain, P. K. (2022). Fabrication of personalized lithophane via additive manufacturing. *Sustainable Operations and Computers*, 3, 17–21. https://doi.org/10.1016/j.susoc.2021.09.001

Suresh Babu, A., Jaivignesh, M., Srinivasan, S., & Sugavaneswaran, M. (2020). Design of functional gradient porous structure and its fabrication using DMLS process. *Materials Today: Proceedings*, 24, 1561–1569. Elsevier Ltd. https://doi.org/10.1016/j.matpr.2020.04.476

Wei, C., Solanki, N. G., Vasoya, J. M., Shah, A. V, & Serajuddin, A. T. M. (2020). Development of 3D printed tablets by fused deposition modeling using polyvinyl alcohol as polymeric matrix for rapid drug release. *Journal of Pharmaceutical Sciences*, 109(4), 1558–1572. https://doi.org/10.1016/j.xphs.2020.01.015

Wilson, J. M., Jones, N., Jin, L., & Shin, Y. C. (2013). Laser deposited coatings of Co-Cr-Mo onto Ti-6Al-4V and SS316L substrates for biomedical applications. *Journal of Biomedical Materials Research - Part B Applied Biomaterials*, 101(7), 1124–1132. https://doi.org/10.1002/jbm.b.32921

Zeidler, H., Klemm, D., Böttger-Hiller, F., Fritsch, S., Le Guen, M. J., & Singamneni, S. (2018). 3D printing of biodegradable parts using renewable biobased materials. *Procedia Manufacturing*, 21, 117–124. https://doi.org/10.1016/J.PROMFG.2018.02.101

Chapter 10

Recent developments in additive manufacturing materials

Nisha Rani, Deepak Gupta, Hemant Yadav, Saurabh Sharma, Dipesh Kumar Mishra, Neeraj Bisht, Arun Kumar Chaudhary, and Alka Verma

10.1 INTRODUCTION

The massive industrial and technological improvements over the past few decades have led to a growing market perception of the necessity for rapid, dependable functional parts that can be supplied to customers [1]. The name "rapid prototyping" (RP) was created to describe this method, which is used in many different industries. Technologies using digital data to create physical goods are commonly referred to as "RP" [2,3]. You can say AM and 3D printing are two different sides of the same coin. Additive manufacturing and 3D printing are interchangeable. The information is transferred from CAD file to the machine, and the material is extruded out from a nozzle layer by layer. It removes material to manufacture a final product. AM has become most popular across various industries. Although the sector tends to use additive manufacturing rather than 3D printing, networking, marketing, and media. AM or 3D printing is known as rapid prototyping. Items made are quicker than using other methods, and rapid prototyping yields better mechanical properties [4]. In 3D printing, for example, rapid prototyping is a technique for creating a model of any structure [5–9]. Applications for additive manufacturing can be found in a number of areas, including manufacturing, aviation, defense, and medicine [10]. Even though 3D printing has many uses, the quality of the final product is examined by the optimization method and degree of experimentation [9]. First, modeling software designs a CAD model, which is subsequently stored in the STL file format. The STL file is separated into layers by slicing software before being sent to AM machinery. After that, the printer starts the 2D-layer developing process, which creates a 3D-model layer-by- layer.

The AM components exhibit remarkable attributes, variety, and flexibility. The distinctive feature of AM techniques is their ability to fabricate some highly special parts such as mold holes with internal cooling tunnels or light hollow shapes. In many instances, substituting traditional production with AM techniques results in cost savings [11]. Due to a significant

 DOI: 10.1201/9781003607052-10

reduction in the design cycle, products can be introduced to the market more rapidly. Metal components with significantly high strength-to-weight ratios are produced without any shape restrictions. Additionally, feature quality and geometric complexity are guaranteed. AM is under green manufacturing techniques, minimizing noise pollution and scrap. Another benefit is the elimination of the requirement for highly specialized workshops. AM is characterized by significant cost savings, through the removal of jigs and fixtures that arise from the lack of tools [12–14]. AM approaches gain additional efficiency when pieces are processed simultaneously or nestled through meticulous layout optimization.

10.2 METHODS IN ADDITIVE MANUFACTURING

Today, numerous AM processes are available commercially.

AM methods have been categorized differently by several researchers. One classification is based upon ASTM-F42 committee, which divides AM into seven different groups. This includes: directed energy deposition (DED), powder bed fusion (PBF), sheet lamination (SL), material extrusion (ME), material jetting (MJ), vat photopolymerization (VP), and binder jetting (BJ).

These categories are arranged according to how much energy they use. For instance, vat photopolymerization uses the least amount (Figure 10.1) of energy (usually a 40–50 mW laser) to cure layers, whereas DED uses a lot of energy (about 4–6 kW).

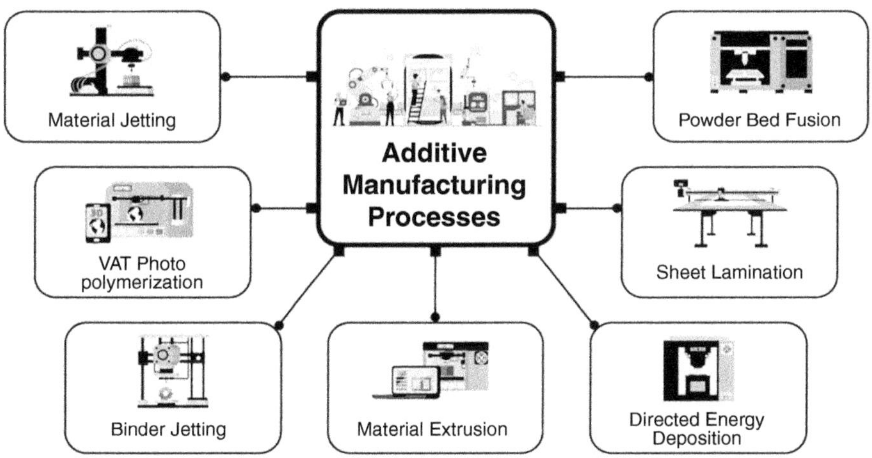

Figure 10.1 Methods of additive manufacturing [15].

10.2.1 Vat photopolymerization

VP is an AM technique that uses light-activated polymerization to selectively cure a liquid photopolymer in a vat. For typically functioning, a light-curable resin is obtained in a container and irradiated with ultraviolet (UV) or visible light [16,17]. Top-down and bottom-up are the two primary system designs for a VP system [18]. The platform is positioned in the resin just under the surface when using a top-down method. After employing light to cure the resin layer above the surface, the stage is brought downcast toward the lowest point of the resin tank to let fresh resin enter and cure the subsequent layer. Resin tank bottom stays transparent while using bottom-up method; in addition, the resin is preserved by light striking under the tank. The platform rises to make space for the new resin to fill the next layer.

Good surface polish, high speed, sufficient accuracy, and other advantages are the key advantages of VP methods. These methods also have certain constraints, like the requirement for support structures, raw material restrictions, and post-processing/curing. VP uses plastic and polymer raw materials, particularly UV-curable photopolymeric resins.

10.2.2 Power bed fusion

Power bed is selectively fused by heat energy in PBF [19]. Lasers, electron beams, or indiscriminate electromagnetic energy are examples of energy sources [20]. PBF methods create 3D components by selectively melting and consolidating build material (powdered form) using high energy power heat source [21–24]. In its simplest form, the shape raised area is covered by an even layer of material of a given thickness. This layer's thickness typically ranges between 30 and 150 mm [25]. The required area of the dropped layer is fused by the temperature source. The construction platform is then lowered, powder is dispersed, and the subsequent layer is fused. Using PBF, similar procedures are done to create 3D objects.

Depending upon the heat source, PBF is additionally known as laser beam primarily based PBF (L-PBF) or electron beam primarily based PBF (E-PBF) [26,27]. They include E-PBF (includes electron beam melting) and L-PBF (includes SLM, direct metal laser sintering, etc.). The main advantages of PBF include their ability to produce extremely complicated components, a large range of material selections, part nesting, especially with polymers and other benefits [28–30]. The spectrum of materials that PBF can treat is rather extensive. Apart from standard materials, PBF can effectively process a specific class of materials, including steels, metals like Inconel and Ti-6Al-4V, thermoplastics, polyamides, and polystyrene [31,32]. PBF processes have certain disadvantages in addition to their advantages, such as the necessity for supports, particularly for metals, high power requirements, post-processing, high cost, prolonged construction, etc.

10.2.3 Material extrusion

The basic concept of extrusion involves attempting to continuously push pressured semi-molten material through a nozzle or at different speeds to create layers following solidification, selectively dispense material through a nozzle, or aperture [33,34]. Temperature or chemical change-based control mechanisms are two possible approaches for layer development [35]. Any extrusion-based approach must follow these fundamental steps: (1) loading material from chamber, (2) liquefy it by pressure and/or heat, (3) extrusion from nozzle, and (4) deposition of layer [36,37]. Extrusion-based AM techniques fall into a number of different categories. The most widely used approach depends on whether or not material melting is taking place. Common methods that include melting materials include fused filament fabrication, fiber deposition, precision extrusion manufacturing/deposition, FDM, and multi-phase jet solidification [38,39]. The main techniques that don't involve melting materials are solvent-based extrusion free forming, pressure-assisted microsyringe, direct-write assembly, bio-plotting, robot casting, and low-temperature deposition [40–42]. Numerous challenges arise when these techniques are used on a large-scale, including in-situ process monitoring, nozzle design, standards gaps, and generally poor surface polish.

10.2.4 Material jetting

Droplets of material are dropped using the MJ- AM method [43]. MJ is also known as Multijet (3D Systems) and Polyjet (Stratasys) [44]. It uses piezo/thermal printing heads for deposition of droplets of liquid photopolymers, which are healed by UV light lamps. During most basic operations, material (build) is injected via printhead at pre-specified locations using either thermal or piezoelectric methods over the build platform [45,46]. The initial layer is formed by these deposited droplets solidifying. To create 3D objects and deposit additional layers, the same procedures are repeated. UV light is used for curing and solidifying layers once they have cooled.

10.2.5 Binder jetting

In BJ methods, powder materials are connected by means of the application of a liquid binding agent [43]. BJ is a type of AM that fuses powder particles together by spraying a liquid binder over a liquid bed. In theory, it is nothing but 2D printing. Output of BJ has 3D objects, in contrast to 2D printing [47,48]. Compared to material jetting, BJ offers comparatively high build speed [49,50]. In its most basic form, a roller is used to spread build material, which is typically in powder form, over the platform. To bind substrate particles together to build a layer, an inkjet printhead sprays binder (usually in liquid form) across the build material, on specified locations. Then, we reduce build platform to create space for the next layer and repeat the whole procedure until the 3D part is formed. The BJ method offers various advantages, a few of which overlap with the material jetting process. It is also noteworthy to mention the high speed, high capacity to print-colored objects, the high compatibility of materials, high build volume, low-to-zero residual tensions and other beneficial factors [51,52]. BJ has certain disadvantages, including porosity defects, shrinkage, and unsuitability for structural components [53,54].

10.2.6 Sheet lamination

In SL, the part is formed by bonding sheets of material [43]. This method produces 3D objects via thin sheets as feedstock materials [39]. To achieve the required shape and size, these thin sheets are layered, laminated, and molded. In order to provide the requisite 3D component, SL approaches can be described as hybrid AM that involves simultaneous material addition and subtraction [55]. Laminated object manufacture (LOM) and ultrasonic consolidation (UC) are two primary variations of SL [56]. The machine receives thin sheets that are first stacked and then sliced with an appropriate energy source (such as a laser beam). Depending upon geometric requirements, the cutting process might take place either before or after sheets are joined. The sheet lamination technique has several desirable features, including lower raw material costs, less requirement for support structures, greater speeds, modest machine setup costs, less thermal stress, fewer twisted, or deformed components, no chemical reaction and ease of producing larger parts [57,58]. This process has some drawbacks, poor surface smoothness, inadequate bonding at interfaces, appropriateness for sheet-type raw materials, lower dimensional accuracy, non-reusable waste material, inability to build hollow components, etc.

10.2.7 Directed energy deposition

A high thermal energy source is used to melt materials while they are being deposited, fusing them together. In its most basic form, DED procedures are comprised of three components: a feedstock, substrate bed with motion controls and a heat source (electron beam, a laser or plasma arc) [59]. A schematic of a typical DED approach is shown in Figure 10.1. First, a molten pool is created using a heat source and filler material, either wire, powder, or a combination of the two is added or injected [60]. As soon as energy source is removed from the deposition site, the molten pool instantly solidifies causing feedstock elements to fuse and unite to produce layer-by-layer structures [61]. Compared to PBF techniques, DED techniques have higher deposition rates, are more repairable, can create functional materials, and more [62–64]. Furthermore, there are a few common disadvantages to DED procedures, which primarily involve the need for a support structure and somewhat poor surface polish [65,66].

10.3 ADDITIVE MANUFACTURING MATERIALS

In AM,manufacturing materials are important. To produce sound parts, materials must be molded into the specified form, possess sufficient qualities, and have appropriate service features. Usually, performance and development of novel applications are made possible by freeform design and optimization (Figures 10.2 and 10.3).

The distribution of materials utilized in AM is shown in Figure 10.1. The classification of materials is shown in Figure 10.2 and some are explained.

10.3.1 Polymers

Polymers in the form of filament, powder, and resin are becoming more popular for AM due to their lower production costs [67]. Benefits including low-maintenance printing systems and energy-efficient materials have grown in popularity in recent years. In material extrusion and AM powder bed fusion, thermoplastic polymers are often used. Semi-crystalline

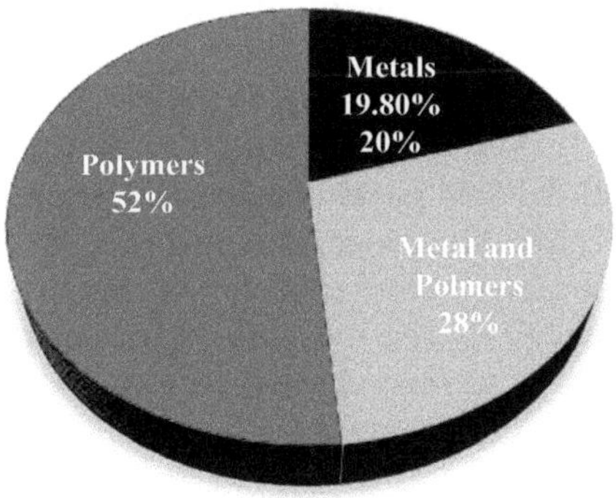

Figure 10.2 Percentage distribution of materials in AM.

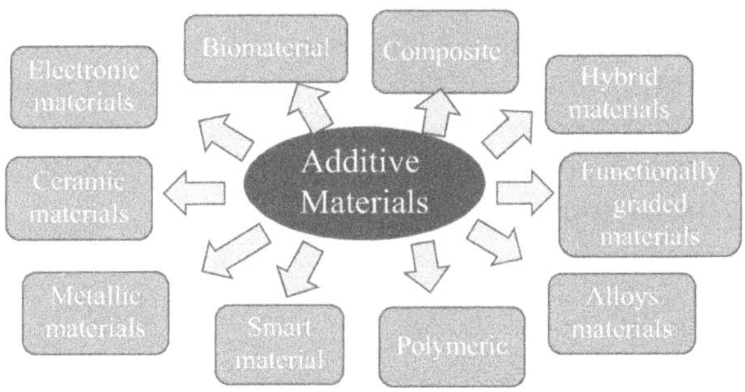

Figure 10.3 Classification of materials in AM.

polymers are used for sintering in powder bed fusion, whereas amorphous thermoplastics are typically taken into consideration for material extrusion (FDM). Material extrusion often employs thermoplastics with amorphous, as well as melting properties. These polymer types, including the most famous of these "ABS" (acrylonitrile butadiene styrene) and "PLA" (polylactic acid), have a softening point at a certain temperature range. This produces a visco-elastic material that can be extruded using a nozzle diameter of 0.2 to 0.5 mm, similar to FDM technology. Among the most common polymers used in additive manufacturing are thermoplastics including PLA, PVA (polyvinyl alcohol), ABS, PEEK (polyetheretherketone) [68], TPU (polyurethane), [69] and nylon [70,71].

10.3.2 PLA

One of the most well-known biopolymers among biodegradable polymers in the market utilized as fixations is polylactide (PLA). "US Food and Drug Administration (FDA)" approved PLA products as biopolymer for clinical usage due to their exceptional mechanical properties, bio-compatibility, biodegradability, and processability [72]. For usage in biodegradability applications, PLA, a biodegradable thermoplastic, has undergone a significant amount of processing. Its bioresorbability does not overwhelm the body or create foreign material, and its biodegradation products (CO_2 and H_2O) are entirely non-toxic [73]. Numerous production factors (input parameters) and their effects on PLA FDM parts' mechanical properties (output parameters) were examined [74]. The process factors of layer thickness, filling percentage, and deposition angle were used to maximize the maximal flexural force in PLA specimens [75]. The mechanical properties of PLA specimens were examined in relation to several input factors, including feed rate, layer thickness, and construction orientation [76].

10.3.3 ABS

In AM, ABS is a distinguished material. This material was the first to be used with commercial 3D printers. It is favored due to its superior mechanical properties and affordability. ABS is known for its exceptional toughness and impact strength, which makes it possible to print 3D parts. Because of its wide range of glass transition temperatures, ABS material can withstand higher temperatures before deforming. As a result, it is best suited for outdoor or applications of high-temperature [77].

10.3.4 PEEK

PEEK, higher melting point, thermal stress, and fractures are more likely to occur. Therefore, for PEEK material to modify its own properties, exact parameter settings are needed [78]. PEEK material's mechanical strengths and thermal deformation were examined using a range of FDM printing parameter settings [79,80]. PEEK can be formed into a filament, and different construction orientations and extrusion pathways can be used to create specimens. Extrusion freeform porous PEEK components' mechanical properties were investigated [81]. PEEK pieces produced by material extrusion with the FDM process were examined for their tensile properties [82]. Ultimately, it was determined that, in comparison to other filaments, PEEK filament exhibits higher mechanical properties. Mechanical properties of 3D-printed objects are essential indicators for determining the machine's print quality.

10.3.5 Reinforced polymer for AM

One major obstacle for fiber-reinforced materials is damage to fibers throughout many pre-processing steps, like filament production and 3D printing. Printing polymers reinforced with continuous carbon fiber is noteworthy. As an example, scientists looked upon 3D printer produced carbon fiber-reinforced thermoplastic (CFRT) [83]. Other researchers created a method for printing CFRT using FFF that same year [84]. Here, PLA served as the parent material, while the reinforcements are either carbon fibers or twisted threads of jute. Thus, in terms of mechanical properties, unidirectional carbon fiber-reinforced thermoplastics fared better than jute and unreinforced thermoplastics.

10.3.6 Metals and alloys

Currently, steels, aluminum, titanium, and nickel alloys are processed and used in various applications with AM [85]. Directed energy deposition and powder bed fusion are two AM processes based on powder-based technologies, and in the industry, they are utilized for manufacturing high-performance metal components. The earliest materials to be treated were steel alloys that are now used across industries that include aerospace and automotive [86]. Additionally, additive manufacturing is seeing growth in the use of pure metals [87]. For load-bearing implant applications, titanium alloys, mainly Ti-6Al-4V, are best available in medicine [88].

High creep and corrosion resistance are necessary for 3D printed components including turbochargers, engine turbine blades, petrochemical equipment, and heat exchangers, made of nickel alloys [89].

Aluminum (Al) and its alloys are strong, corrosion-resistant, lightweight, and simple to weld. They are perfect for a wide range of applications like construction, aerospace, automotive, and defense. Among the several alloy combinations employed in the SLM technique, aluminum-silicon-based alloys have been frequently used due to their superior fabricability [90].

10.3.7 Metal-polymer hybrid

The fact that thermoplastic 3DP alone does not have thermal, mechanical, or electrical qualities that an industry requires is one of its main problems [91]. Lot of interest in finding a practical method to achieve metal 3DP. Although 3D printing with thermoplastics materials is well advanced and quickly and affordable in producing complex topologies, 3D printing with metals is still challenging due to monetary constraints. Bulk metallic glasses (BMGs), class of metallic materials, behave similar to thermoplastics when heated.

10.3.8 Ceramics

The inability of certain raw materials, like as concrete and ceramics, to fuse together when heated to their melting temperatures results in their limited application in AM [92]. Alternatively, metals and polymers sufficiently fuse at their melting points. A significant obstacle prior to AM technologies was that ceramics had significantly higher melting temperatures than metals and polymers. Therefore, AM procedures that can create ceramic parts give them mechanical qualities that are on pace with those produced using traditional manufacturing methods [93–96]. Although PBF-based AM procedures are particularly effective and cost-effective ways to create ceramic parts, there are still issues with the initial raw material availability for getting feedstock. PBF is frequently used in the processing of ceramics, including calcium phosphate, silicon carbide, silica, etc. [25]. Ceramic materials are used in various industries, including electronics, microchips, automotive (for the development of engine and propulsion components), and aerospace [97]. The use of ceramic AM in tissue engineering and biomaterials is among its most promising uses. Ceramics take additional time and post-processing, which adds to the process's cost. This does not, however, decrease the attraction of these AM methods for producing components with complex shapes. Scaffolds constructed of ceramic for teeth and bones are a highly unique example of ceramic AM parts. Given how quick the process is compared to more conventional methods like, sintering or casting, this is a crucial component of tissue engineering.

10.3.9 Smart materials

A family of sophisticated materials known as "smart materials" are able to change their characteristics in a predictable and reversible way in response to external stimuli like light, temperature, magnetic, or electric fields, pH, and mechanical force. Across industries, the combination of smart materials and additive manufacturing (AM) is opening up new opportunities for the development of intelligent, adaptable, and functioning systems.

Smart materials can be processed layer-by-layer in additive manufacturing (AM) to create multifunctional components and complicated geometries that were previously impossible to achieve with conventional manufacturing techniques. The following are important types of smart materials utilized in AM:

10.3.9.1 Shape memory alloys (SMAs)

When subjected to particular stimuli, such as heat, these materials can regain their original shape after deformation. SMA components are fabricated using AM processes such as selective laser melting (SLM) for usage in actuators, biomedical devices, and aircraft.

10.3.9.2 Piezoelectric materials

These respond to mechanical stresses by producing an electric charge. The manufacturing of sensors, energy harvesters, and adaptable structures is made possible by their use in AM.

10.3.9.3 Hydrogels

Depending on the temperature or pH of the surrounding environment, hydrogels can expand or contract. Soft robotics, medication delivery systems, and tissue engineering have all benefited from the advancements made possible by 3D printing hydrogels.

10.3.9.4 Self-healing polymers

These substances fix damage on their own, prolonging component life. AM makes it possible to precisely structure self-healing networks for use in electronics, infrastructure, and aerospace.

10.3.9.5 Photomechanical materials

These react to light by changing their optical characteristics or shape. They are being utilized more and more in 3D printing, a branch of AM in which, under particular circumstances, printed things change over time.

Innovations like responsive wearables, adaptive structures, and customized medical equipment have been made possible by the combination of AM and smart materials. Nevertheless, there are still issues to be resolved, such as integrating intelligent features into intricate systems, attaining consistent performance, and optimizing material compositions for printability. Enhancing AM methods' compatibility with smart materials and investigating novel materials that could broaden the scope of intelligent manufacturing are the main goals of ongoing research.

10.3.10 Functionally graded materials

Advanced materials known as functionally graded materials (FGMs) exhibit a slow change in composition, structure, and characteristics throughout their volume. By purposefully combining the beneficial qualities of various materials, this gradient improves performance and functionality. Applications needing specific thermal, mechanical, or chemical characteristics, like energy systems, biomedical implants, and aerospace components, are especially well-suited for FGMs. Additive manufacturing (AM) technologies, which provide fine-grained control over material composition and deposition at the voxel level, have greatly enhanced the creation of FGMs. FGMs with spatially variable material properties have been created using methods such as binder jetting, selective laser melting, and directed energy deposition (DED). FGMs with a metal-to-ceramic gradient, for example, can have great strength on one side and outstanding thermal resistance on the other, which makes them perfect for wear-resistant surfaces and thermal barrier coatings.

10.3.11 Composite materials

In additive manufacturing (AM), composite materials that combine two or more components to produce a material with improved properties have drawn a lot of interest. Composites can achieve high strength-to-weight ratios, higher electrical conductivity, improved thermal resistance, and greater wear resistance by utilizing the distinct qualities of many materials. The fabrication of high-performance, multi-purpose parts for consumer, automotive, healthcare, and aerospace applications is made possible by the use of composite materials in additive manufacturing. Types of composites in AM are given as follows.

10.3.11.1 Polymer matrix composites (PMCs)

These are made of a polymer matrix that has been strengthened with fibers or other particles. PMCs reinforced with materials like carbon fibers, glass fibers, or nanomaterials (like graphene) are printed using additive processes like fused deposition modeling (FDM) or vat photopolymerization. These composites are perfect for lightweight structural components because of their high rigidity and lightweight properties.

10.3.11.2 Metal matrix composites (MMCs)

MMCs enhance strength, wear resistance, and thermal stability by incorporating ceramic or metallic reinforcements into a metallic matrix. MMCs for high-performance applications, such as aerospace engine components, are frequently fabricated by directed energy deposition (DED) and powder bed fusion (PBF).

10.3.11.3 Carbon or ceramic fibers

These are frequently used to strengthen ceramic materials, which serve as the matrix in ceramic matrix composites (CMCs). These composites withstand high-temperature settings due to their remarkable chemical and thermal resistance. CMC is produced using additive methods like as stereolithography and binder jetting.

10.3.11.4 Hybrid composites

To attain specific qualities, hybrid composites blend various reinforcing types or matrix materials. By precisely distributing reinforcements, additive manufacturing makes it possible to create multi-material or functionally graded composites.

10.3.12 Biomaterials

Biomaterials are substances that interact with biological systems for medical purposes, as tissue engineering, prosthetics, implants, and drug delivery systems. Additive manufacturing (AM) or 3D printing, revolutionized the use of biomaterials. It enables the fabrication of patient-specific, complex, and functional structures with precise control over material distribution and architecture.

10.4 CHALLENGES IN AM

The study of AM has advanced significantly in the modern era, and its applications and developments have followed suit. To enable AM's usability, TL4 level technology facilitate its transition from research labs towards actual companies, a number of obstacles must be fully addressed. One main problem is compatibility of raw materials with AM techniques. The limitations of AM materials include dimensional accuracy, surface smoothness of manufactured parts, anisotropic behavior, etc. Absence of testing facilities of certain parts and materials made via AM method is still another difficulty.

The most significant obstacle is the prevalence of many flaws in items made using additive manufacturing techniques. The following subsection discusses a few of the prevalent flaws:

10.4.1 Balling phenomena

Balling phenomenon, also known as bead up. It is a flaw in which liquid material does not moisten the underlying surfaces, causing a beaded scan track to form. This increases surface roughness and the likelihood of pores forming [98]. This flaw mostly affects AM methods that rely on laser sintering.

10.4.2 Porosity

Without the assistance of outside forces, the majority of binding mechanisms are linked to temperature changes under gravity and capillary action. Repeated keyhole emergence, gas trapping that creates small gaps during atomization, insufficient substrate penetration of subsequent layers, etc. are some of the reasons of porosity/void defects [99]. The size of keyhole pores is less than 100 microns.

10.4.3 Cracks

As the metals melt and solidify quickly during fusion-based AM methods, fractures are another frequent problem in parts made using these processes [100]. Substantial temperature gradients and similarly substantial thermal residual strains are generated as a result of this rapid cooling. Crack initiation is caused by a large residual stress and a strong temperature gradient. Mushy zones are exposed to tensile pressures that cause cracking to begin at liquid films, liquation cracking [101] is a frequent problem in these areas. It is also possible to understand additional fissures. Layers will mutually separate if the residual tension at the interface exceeds the alloy's yield strength.

10.4.4 Distortion

When material experiences stress development due to volume shrinking, AM parts produce distortion defects.

10.4.5 Inferior surfaces

A common flaw in AM parts is the inferior surface, which is mostly caused by the staircase effect. Some other contributing elements includes orientation of the part, use of aged material (SLS), rough bead deposition (FDM), surface tension and balling (PBF), semi-molten powder (SLM), and restricted tool precision (EBM). If all of these process-specific obstacles are resolved, better surfaces can be produced. Attempts to minimize surface roughness have been found to result in greater manufacturing costs and poorer production rates. This is primarily due to the increasing need for post-processing and processing time.

10.5 CONCLUSION

In additive manufacturing (AM), materials are crucial, especially when it comes to materials for designed structural applications. Materials must have acceptable service properties, be shaped into the right feedstock, and possess the right qualities for processing, in particular AM fabricator. There are seven main categories of additive manufacturing (AM) processes, which are: BJ, ME, DED, MJ, SL, PBF, and VP; and for each category, there are materials that can be used, being optimized for their use in AM applications. Metals, ceramics, composites, and polymers are covered. A review of AM binding mechanisms is given. Strength, ductility, stiffness, toughness, optical qualities, fatigue, and electrical qualities are among the service characteristics of AM materials. Currently, efforts are concentrated on identifying and removing the causes of problems in AM parts. These consist of micro-structural effects, interfaces, binding flaws, and porosity.

REFERENCES

1. Rejeski, D., Zhao, F., & Huang, Y. (2018). Research needs and recommendations on environmental implications of additive manufacturing. *Additive Manufacturing*, 19, 21–28.
2. Matta, A. K., Raju, D. R., & Suman, K. N. S. (2015). The integration of CAD/CAM and rapid prototyping in product development: A review. *Materials Today: Proceedings*, 2(4–5), 3438–3445.
3. Eyers, D. R., & Potter, A. T. (2017). Industrial additive manufacturing: A manufacturing systems perspective. *Computers in Industry*, 92, 208–218.
4. Serra, T., Planell, J. A., & Navarro, M. (2013). High-resolution PLA-based composite scaffolds via 3-D printing technology. *Acta Biomaterialia*, 9(3), 5521–5530.
5. Sathish, T., Vijayakumar, M. D., & Ayyangar, A. K. (2018). Design and fabrication of industrial components using 3D printing. *Materials Today: Proceedings*, 5(6), 14489–14498.
6. Putame, G., Terzini, M., Carbonaro, D., Pisani, G., Serino, G., Di Meglio, F., Castaldo, C., & Massai, D. (2019). Application of 3D printing technology for design and manufacturing of customized components for a mechanical stretching bioreactor. *Journal of Healthcare Engineering*, 2019(1), 3957931.
7. Vaezi, M., Seitz, H., & Yang, S. (2013). A review on 3D micro-additive manufacturing technologies. *The International Journal of Advanced Manufacturing Technology*, 67, 1721–1754.
8. Lee, D., Kim, H., Sim, J., Lee, D., Cho, H., & Hong, D. (2019). Trends in 3D printing technology for construction automation using text mining. *International Journal of Precision Engineering and Manufacturing*, 20, 871–882.
9. Macdonald, E., Salas, R., Espalin, D., Perez, M., Aguilera, E., Muse, D., & Wicker, R. B. (2014). 3D printing for the rapid prototyping of structural electronics. *IEEE Access*, 2, 234–242.
10. Sheoran, A. J., & Kumar, H. (2020). Fused deposition modeling process parameters optimization and effect on mechanical properties and part quality: Review and reflection on present research. *Materials Today: Proceedings*, 21, 1659–1672.
11. Srivastava, M., Rathee, S., Maheshwari, S., & Kundra, T. K. (2019). *Additive Manufacturing: Fundamentals and Advancements*. Boca Raton, FL: CRC Press.

12. Thomas, D. (2016). Costs, benefits, and adoption of additive manufacturing: A supply chain perspective. *The International Journal of Advanced Manufacturing Technology*, 85, 1857–1876.
13. Baumers, M., Tuck, C., Wildman, R., Ashcroft, I., Rosamond, E., & Hague, R. (2013). Transparency built-in: Energy consumption and cost estimation for additive manufacturing. *Journal of Industrial Ecology*, 17(3), 418–431.
14. Raja, V., Zhang, S., Garside, J., Ryall, C., & Wimpenny, D. (2006). Rapid and cost-effective manufacturing of high-integrity aerospace components. *The International Journal of Advanced Manufacturing Technology*, 27, 759–773.
15. Mobarak, M. H., Islam, M. A., Hossain, N., Al Mahmud, M. Z., Rayhan, M. T., Nishi, N. J., & Chowdhury, M. A. (2023). Recent advances of additive manufacturing in implant fabrication–A review. *Applied Surface Science Advances*, 18, 100462.
16. Zhang, F., Zhu, L., Li, Z., Wang, S., Shi, J., Tang, W., ... & Yang, J. (2021). The recent development of vat photopolymerization: A review. *Additive Manufacturing*, 48, 102423.
17. Miao, G., & Ma, C. (2019). Vat photopolymerization 3D printing of ceramics: Effects of fine powder. *Manufacturing Letters*, 21, 20–23.
18. Medellin, A., Du, W., Miao, G., Zou, J., Pei, Z., & Ma, C. (2019). Vat photopolymerization 3d printing of nanocomposites: A literature review. *Journal of Micro-and Nano-Manufacturing*, 7(3), 031006.
19. ASTM, A. (2012). *F2792-12a: Standard Terminology for Additive Manufacturing Technologies*. West Conshohocken, PA: ASTM International.
20. Bourell, D. L., & Wohlers, T. (2020). *Introduction to additive Manufacturing*. Fort Collins, CO: Wohlers Associates.
21. Gu, D. (2015). Novel Ti-based nanocomposites by selective laser melting (SLM) additive manufacturing (AM): Tailored nanostructure and performance. In *Laser Additive Manufacturing of High-Performance Materials* (pp. 73–113). Cham: Springer.
22. Brandt, M. (2017). The role of lasers in additive manufacturing. In *Laser Additive Manufacturing* (pp. 1–18). Cambridge: Woodhead Publishing (Elsevier).
23. Kumar, S. (2003). Selective laser sintering: A qualitative and objective approach. *JOM*, 55, 43–47.
24. Verma, R., & Kaushal, G. (2019). State of the art of powder bed fusion additive manufacturing: A review. In *3D Printing and Additive Manufacturing Technologies* (pp. 269–279). Boca Raton, FL: CRC Press (Taylor & Francis).
25. Avrampos, P., & Vosniakos, G. C. (2022). A review of powder deposition in additive manufacturing by powder bed fusion. *Journal of Manufacturing Processes*, 74, 332–352.
26. Qu, M., Guo, Q., Escano, L. I., Nabaa, A., Hojjatzadeh, S. M. H., Young, Z. A., & Chen, L. (2022). Controlling process instability for defect lean metal additive manufacturing. *Nature Communications*, 13(1), 1079.
27. Lewin, S., Fleps, I., Åberg, J., Ferguson, S. J., Engqvist, H., Öhman-Mägi, C., ... & Persson, C. (2021). Additively manufactured mesh-type titanium structures for cranial implants: E-PBF vs. L-PBF. *Materials & Design*, 197, 109207.

28. Goodridge, R., & Ziegelmeier, S. (2017). Powder bed fusion of polymers. In *Laser Additive Manufacturing* (pp. 181–204). Cambridge: Woodhead Publishing (Elsevier).
29. Kruth, J. P., Mercelis, P., Van Vaerenbergh, J., Froyen, L., & Rombouts, M. (2005). Binding mechanisms in selective laser sintering and selective laser melting. *Rapid Prototyping Journal*, 11(1), 26–36.
30. Shirazi, S. F. S., Gharehkhani, S., Mehrali, M., Yarmand, H., Metselaar, H. S. C., Kadri, N. A., & Osman, N. A. A. (2015). A review on powder-based additive manufacturing for tissue engineering: Selective laser sintering and inkjet 3D printing. *Science and Technology of Advanced Materials*, 16(3), 033502.
31. Neikter, M., Åkerfeldt, P., Pederson, R., & Antti, M. L. (2017, October). Microstructure characterisation of Ti-6Al-4V from different additive manufacturing processes. In *IOP Conference Series: Materials Science and Engineering* (Vol. 258, No. 1, p. 012007). Bristol: IOP Publishing.
32. Nie, X., Chen, Z., Qi, Y., Zhang, H., Zhang, C., Xiao, Z., & Zhu, H. (2020). Effect of defocusing distance on laser powder bed fusion of high strength Al–Cu–Mg–Mn alloy. *Virtual and Physical Prototyping*, 15(3), 325–339.
33. Suwanpreecha, C., & Manonukul, A. (2022). A review on material extrusion additive manufacturing of metal and how it compares with metal injection moulding. *Metals*, 12(3), 429.
34. Rathee, S., & Srivastava, M. (2021). Layout optimization for FDM process by multi-objective optimization using RSM and GRA. In *Fused Deposition Modeling Based 3D Printing* (pp. 505–515). Boca Raton, FL: CRC Press (Taylor & Francis).
35. Oleff, A., Küster, B., Stonis, M., & Overmeyer, L. (2021). Process monitoring for material extrusion additive manufacturing: A state-of-the-art review. *Progress in Additive Manufacturing*, 6(4), 705–730.
36. Srivastava, M., Maheshwari, S., Kundra, T. K., Rathee, S., & Yashaswi, R. (2016). Experimental investigation of process parameters for build time estimation in FDM process using RSM technique. In *CAD/CAM, Robotics and Factories of the Future: Proceedings of the 28th International Conference on CARs & FoF 2016* (pp. 229–241). New Delhi: Springer India.
37. Spoerk, M., Holzer, C., & Gonzalez-Gutierrez, J. (2020). Material extrusion-based additive manufacturing of polypropylene: A review on how to improve dimensional inaccuracy and warpage. *Journal of Applied Polymer Science*, 137(12), 48545.
38. Casini, M. (2022). Advanced building construction methods. *Construction*, 4, 405–470.
39. Zhang, X., & Liou, F. (2021). Introduction to additive manufacturing. In *Additive Manufacturing* (pp. 1–31). Cambridge: Elsevier (Woodhead Publishing).
40. Paterlini, A., Le Grill, S., Brouillet, F., Combes, C., Grossin, D., & Bertrand, G. (2021). Robocasting of self-setting bioceramics: From paste formulation to 3D part characteristics. *Open Ceramics*, 5, 100070.
41. Bevis, J. B., Dunlavey, S., & Martinez-Duarte, R. (2021). Comparing the performance of different extruders in the Robocasting of biopolymer-nanoparticle composites towards the fabrication of complex geometries of porous Tungsten Carbide. *Procedia Manufacturing*, 53, 338–342.

42. Lu, X., Lee, Y., Yang, S., Hao, Y., Evans, J. R., & Parini, C. G. (2010). Solvent-based paste extrusion solid freeforming. *Journal of the European Ceramic Society*, 30(1), 1–10.
43. García-Domínguez, A., Claver, J., Camacho, A. M., & Sebastián, M. A. (2020). Analysis of general and specific standardization developments in additive manufacturing from a materials and technological approach. *IEEE Access*, 8, 125056–125075.
44. Mora, S., Pugno, N. M., & Misseroni, D. (2022). 3D printed architected lattice structures by material jetting. *Materials Today*, 59, 107–132.
45. Kaindl, R., Homola, T., Rastelli, A., Schwarz, A., Tarre, A., Kopp, D., ... & Waldhauser, W. (2022). Atomic layer deposition of oxide coatings on porous metal and polymer structures fabricated by additive manufacturing methods (laser-based powder bed fusion, material extrusion, material jetting). *Surfaces and Interfaces*, 34, 102361.
46. Bezek, L. B., Chatham, C. A., Dillard, D. A., & Williams, C. B. (2022). Mechanical properties of tissue-mimicking composites formed by material jetting additive manufacturing. *Journal of the Mechanical Behavior of Biomedical Materials*, 125, 104938.
47. Izdebska-Podsiadły, J. (2022). Classification of 3D printing methods. In *Polymers for 3D Printing* (pp. 23–34). Cambridge, MA: William Andrew Publishing (Elsevier).
48. Caballero, F. G. (2021). *Encyclopedia of materials: Metals and alloys*. Oxford: Elsevier.
49. Bai, Y., & Williams, C. B. (2015). An exploration of binder jetting of copper. *Rapid Prototyping Journal*, 21(2), 177–185.
50. Ziaee, M., & Crane, N. B. (2019). Binder jetting: A review of process, materials, and methods. *Additive Manufacturing*, 28, 781–801.
51. Miyanaji, H., Orth, M., Akbar, J. M., & Yang, L. (2018). Process development for green part printing using binder jetting additive manufacturing. *Frontiers of Mechanical Engineering*, 13, 504–512.
52. Miyanaji, H., Ma, D., Atwater, M. A., Darling, K. A., Hammond, V. H., & Williams, C. B. (2020). Binder jetting additive manufacturing of copper foam structures. *Additive Manufacturing*, 32, 100960.
53. Dini, F., Ghaffari, S. A., Jafar, J., Hamidreza, R., & Marjan, S. (2020). A review of binder jet process parameters; powder, binder, printing and sintering condition. *Metal Powder Report*, 75(2), 95–100.
54. Li, M., Huang, J., Fang, A., Mansoor, B., Pei, Z., & Ma, C. (2021). Binder jetting additive manufacturing of copper/diamond composites: An experimental study. *Journal of Manufacturing Processes*, 70, 205–213.
55. Rathee, S., Srivastava, M., Pandey, P. M., Mahawar, A., & Shukla, S. (2021). Metal additive manufacturing using friction stir engineering: A review on microstructural evolution, tooling and design strategies. *CIRP Journal of Manufacturing Science and Technology*, 35, 560–588.
56. Bournias-Varotsis, A., Friel, R. J., Harris, R. A., & Engstrøm, D. S. (2018). Ultrasonic Additive Manufacturing as a form-then-bond process for embedding electronic circuitry into a metal matrix. *Journal of Manufacturing Processes*, 32, 664–675.
57. Hehr, A., & Norfolk, M. (2019). A comprehensive review of ultrasonic additive manufacturing. *Rapid Prototyping Journal*, 26(3), 445–458.

58. Mekonnen, B. G., Bright, G., & Walker, A. (2016, January). A study on state of the art technology of laminated object manufacturing (LOM). In *CAD/CAM, Robotics and Factories of the Future: Proceedings of the 28th International Conference on CARs & FoF 2016* (pp. 207–216). New Delhi: Springer India.
59. Gibson, I., Rosen, D. W., Stucker, B., Gibson, I., Rosen, D. W., & Stucker, B. (2010). Sheet lamination processes. In *Additive Manufacturing Technologies: Rapid Prototyping to Direct Digital Manufacturing* (pp. 223–252). New York, NY: Springer.
60. Ahn, D. G. (2021). Directed energy deposition (DED) process: State of the art. *International Journal of Precision Engineering and Manufacturing-Green Technology*, 8(2), 703–742.
61. Singh, A., Kapil, S., & Das, M. (2020). A comprehensive review of the methods and mechanisms for powder feedstock handling in directed energy deposition. *Additive Manufacturing*, 35, 101388.
62. Dávila, J. L., Neto, P. I., Noritomi, P. Y., Coelho, R. T., & da Silva, J. V. L. (2020). Hybrid manufacturing: A review of the synergy between directed energy deposition and subtractive processes. *The International Journal of Advanced Manufacturing Technology*, 110, 3377–3390.
63. Tang, Z. J., Liu, W. W., Wang, Y. W., Saleheen, K. M., Liu, Z. C., Peng, S. T., ... & Zhang, H. C. (2020). A review on in situ monitoring technology for directed energy deposition of metals. *The International Journal of Advanced Manufacturing Technology*, 108, 3437–3463.
64. Zhang, D., Lim, W. Y. S., Duran, S. S. F., Loh, X. J., & Suwardi, A. (2022). Additive manufacturing of thermoelectrics: Emerging trends and outlook. *ACS Energy Letters*, 7(2), 720–735.
65. Li, S. H., Kumar, P., Chandra, S., & Ramamurty, U. (2023). Directed energy deposition of metals: Processing, microstructures, and mechanical properties. *International Materials Reviews*, 68(6), 605–647.
66. Lim, J. S., Oh, W. J., Lee, C. M., & Kim, D. H. (2021). Selection of effective manufacturing conditions for directed energy deposition process using machine learning methods. *Scientific Reports*, 11(1), 24169.
67. Nath, S. D., & Nilufar, S. (2020). An overview of additive manufacturing of polymers and associated composites. *Polymers*, 12(11), 2719.
68. Ni, F., Wang, G., & Zhao, H. (2017). Fabrication of water-soluble poly (vinyl alcohol)-based composites with improved thermal behavior for potential three-dimensional printing application. *Journal of Applied Polymer Science*, 134(24).
69. Han, Y., & Kim, J. (2018). A study on the mechanical properties of knit fabric using 3D printing-Focused on PLA, TPU Filament. *Journal of Fashion Business*, 22(4), 93–105.
70. Xiao, J., & Gao, Y. (2017). The manufacture of 3D printing of medical grade TPU. *Progress in Additive Manufacturing*, 2, 117–123.
71. Amado Becker, A. F. (2016). Characterization and prediction of SLS processability of polymer powders with respect to powder flow and part warpage (Doctoral dissertation, ETH Zurich).
72. Xiao, L., Wang, B., Yang, G., & Gauthier, M. (2012). Poly (lactic acid)-based biomaterials: Synthesis, modification and applications. *Biomedical Science, Engineering and Technology*, 11, 247–282.

73. Aworinde, A. K., Taiwo, O. O., Adeosun, S. O., Akinlabi, E. T., Jonathan, H., Olayemi, O. A., & Joseph, O. O. (2021). Biodegradation profiles of chitin, chitosan and titanium reinforced polylactide biocomposites as scaffolds in bone tissue engineering. *Arab Journal of Basic and Applied Sciences*, 28(1), 351–359.
74. Torres, J., Cole, M., Owji, A., DeMastry, Z., & Gordon, A. P. (2016). An approach for mechanical property optimization of fused deposition modeling with polylactic acid via design of experiments. *Rapid Prototyping Journal*, 22(2), 387–404.
75. Pfeifer, T., Koch, C., Van Hulle, L., Capote, G. M., & Rudolph, N. (2016). Optimization of the FDM additive manufacturing process. *Proceedings of the SPE ANTEC™ Indianapolis*, Indianapolisw, IN, 22–25.
76. Chacón, J. M., Caminero, M. A., García-Plaza, E., & Núnez, P. J. (2017). Additive manufacturing of PLA structures using fused deposition modelling: Effect of process parameters on mechanical properties and their optimal selection. *Materials & Design*, 124, 143–157.
77. Żur, P., Kołodziej, A., Baier, A., & Kokot, G. (2020). Optimization of Abs 3D-printing method and parameters. *European Journal of Engineering Science and Technology*, 3(1), 44–51.
78. Salem Bala, A., bin Wahab, M. S., & binti Ahmad, M. (2016, May). Elements and materials improve the FDM products: A review. In *Advanced Engineering Forum* (Vol. 16, pp. 33–51). Uetikon-Zurich: Trans Tech Publications Ltd.
79. Wu, W. Z., Geng, P., Zhao, J., Zhang, Y., Rosen, D. W., & Zhang, H. B. (2014). Manufacture and thermal deformation analysis of semicrystalline polymer polyether ether ketone by 3D printing. *Materials Research Innovations*, 18(sup5), S5–12.
80. Wu, W., Geng, P., Li, G., Zhao, D., Zhang, H., & Zhao, J. (2015). Influence of layer thickness and raster angle on the mechanical properties of 3D-printed PEEK and a comparative mechanical study between PEEK and ABS. *Materials*, 8(9), 5834–5846.
81. Vaezi, M., & Yang, S. (2015). Extrusion-based additive manufacturing of PEEK for biomedical applications. *Virtual and Physical Prototyping*, 10(3), 123–135.
82. Cicala, G., Latteri, A., Del Curto, B., Lo Russo, A., Recca, G., & Farè, S. (2017). Engineering thermoplastics for additive manufacturing: A critical perspective with experimental evidence to support functional applications. *Journal of Applied Biomaterials & Functional Materials*, 15(1), 10–18.
83. Van Der Klift, F., Koga, Y., Todoroki, A., Ueda, M., Hirano, Y., & Matsuzaki, R. (2015). 3D printing of continuous carbon fibre reinforced thermo-plastic (CFRTP) tensile test specimens. *Open Journal of Composite Materials*, 6(1), 18–27.
84. Matsuzaki, R., Ueda, M., Namiki, M., Jeong, T. K., Asahara, H., Horiguchi, K., ... & Hirano, Y. (2016). Three-dimensional printing of continuous-fiber composites by in-nozzle impregnation. *Scientific Reports*, 6(1), 23058.
85. Frazier, W. E. (2014). Metal additive manufacturing: A review. *Journal of Materials Engineering and Performance*, 23, 1917–1928.
86. Bajaj, P., Hariharan, A., Kini, A., Kürnsteiner, P., Raabe, D., & Jägle, E. A. (2020). Steels in additive manufacturing: A review of their microstructure and properties. *Materials Science and Engineering: A*, 772, 138633.
87. Aboulkhair, N. T., Simonelli, M., Parry, L., Ashcroft, I., Tuck, C., & Hague, R. (2019). 3D printing of aluminium alloys: Additive manufacturing of aluminium alloys using selective laser melting. *Progress in Materials Science*, 106, 100578.

88. Ataee, A., Li, Y., Song, G., & Wen, C. (2017). Metal scaffolds processed by electron beam melting for biomedical applications. In *Metallic Foam Bone* (pp. 83–110). Cambridge: Woodhead Publishing.
89. Graybill, B., Li, M., Malawey, D., Ma, C., Alvarado-Orozco, J. M., & Martinez-Franco, E. (2018, June). Additive manufacturing of nickel-based superalloys. In *Proceedings of the ASME International Manufacturing Science and Engineering Conference (MSEC 2018)*, College Station, Texas, USA. New York, NY: American Society of Mechanical Engineers.
90. Ponnusamy, P., Rahman Rashid, R. A., Masood, S. H., Ruan, D., & Palanisamy, S. (2020). Mechanical properties of SLM-printed aluminium alloys: A review. *Materials*, 13(19), 4301.
91. Pacurar, R., Pacurar, A., & Pop, S. (2018). Designing of an innovative extrusion system for metallic parts made by desktop 3D printing method. In *MATEC Web of Conferences* (Vol. 178, p. 02009). Les Ulis: EDP Sciences.
92. Lee, J. Y., An, J., & Chua, C. K. (2017). Fundamentals and applications of 3D printing for novel materials. *Applied Materials Today*, 7, 120–133.
93. Niu, F., Wu, D., Ma, G., & Zhang, B. (2015). Additive manufacturing of ceramic structures by laser engineered net shaping. *Chinese Journal of Mechanical Engineering*, 28(6), 1117–1122.
94. Kirihara, S. (2016). Additive manufacturing of ceramic components using laser scanning stereolithography. *Welding in the World*, 60, 697–702.
95. Promakhov, V. V. E., Savinykh, A. S., Dubkova, Y. A., Schulz, N. A., Grunt, N. V., & Razorenov, S. V. (2018). Strength properties of aluminum-oxide ceramics prepared by the additive manufacturing method under shock-wave loading. *Technical Physics Letters*, 44, 898–901.
96. Travitzky, N., Bonet, A., Dermeik, B., Fey, T., Filbert-Demut, I., Schlier, L., ... & Greil, P. (2014). Additive manufacturing of ceramic-based materials. *Advanced Engineering Materials*, 16(6), 729–754.
97. Yang, L., & Miyanaji, H. (2017). Ceramic additive manufacturing: A review of current status and challenges. In *Additive Manufacturing* (vol. 21, pp. 710–731). Amsterdam: Elsevier.
98. Mostafaei, A., Zhao, C., He, Y., Ghiaasiaan, S. R., Shi, B., Shao, S., ... & Rollett, A. D. (2022). Defects and anomalies in powder bed fusion metal additive manufacturing. *Current Opinion in Solid State and Materials Science*, 26(2), 100974.
99. Aboulkhair, N. T., Everitt, N. M., Ashcroft, I., & Tuck, C. (2014). Reducing porosity in AlSi10Mg parts processed by selective laser melting. *Additive Manufacturing*, 1, 77–86.
100. Sun, Z., Ma, Y., Ponge, D., Zaefferer, S., Jägle, E. A., Gault, B., ... & Raabe, D. (2022). Thermodynamics-guided alloy and process design for additive manufacturing. *Nature Communications*, 13(1), 4361.
101. Forster, A. M. (2015). Materials testing standards for additive manufacturing of polymer materials: State of the art and standards applicability. *Polymers*, 7(10), 2281–2299.

Chapter 11

Additive manufacturing for sustainable dentistry

Global trends and applications

Ankit Tyagi, Siddharth, Sachin Sirohi, Rekha Nair, Ankita Dadhich, and Abhishek Dadhich

11.1 INTRODUCTION

11.1.1 Toward sustainable digital manufacturing at a global level

The manufacturing world is undergoing a paradigm shift through digitalization, sustainability imperatives, and cyber–physical integration. Additive manufacturing (AM) is now considered an important pioneer of this shift as it allows layer-by-layer fabrication directly from digital models to reduce material wastage, remove tooling, and support complex geometries that were not possible using conventional manufacturing pathways [1]. AM on the global stage as an industrial development strategy is gaining recognition and attention on an industrial scale, especially in emerging economies looking to overcome resource-intensive legacy manufacturing systems [2]. Industry 4.0 and Industry 5.0 paradigms are helping drive the adoption of AM technology at a fast pace that includes intelligent automation, human-inclusive manufacturing, and the responsibility of sustainable production [3,4]. These factors have established AM as not just a production technology, but rather a system level technology that adapts value chains for different types of use, reduce the reliance on logistics, and support distributed manufacturing systems [5] (Figure 11.1).

11.1.2 The sustainability challenges of conventional dental manufacturing

Dentistry is a one-stop shop at the crossroads between healthcare and manufacturing, it has high patient volumes, constant device replacements, and a labor-intensive laboratory workflow. Subtractive and manual processes, such as milling, casting, pressing, and molding, are the dominant manufacturing methods in conventional dental manufacturing, which are inherently material-inefficient and labor-intensive [6]. In ceramic and metallic restorations, any subtractive milling process produces material losses of more than 70–90% of the original blank with particular reference to zirconia- and

 DOI: 10.1201/9781003607052-11

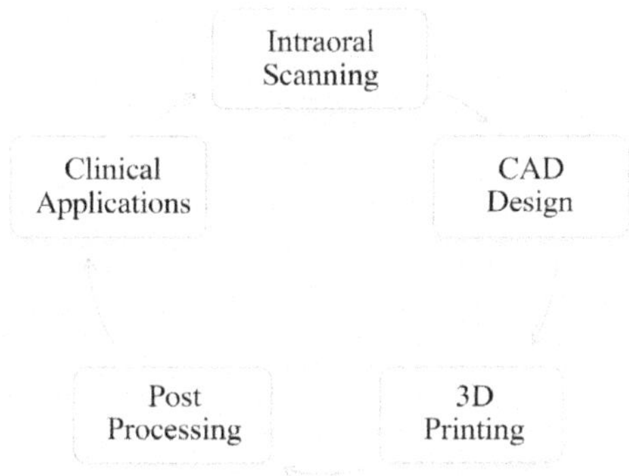

Figure 11.1 Additive manufacturing in dentistry workflow.

titanium-based parts [7]. Besides material wastage, centralized dental laboratory models pose a major threat to sustainability due to recurrent distribution of impressions, models, and restorations between clinics and laboratories, leading to higher carbon emissions and turnaround times [8]. Conventional dental manufacturing operations have an environmental impact: energy-intensive sintering, casting, and thermal post-processing processes, as well as the application of chemical cleaning and finishing agents [9,10].

11.1.3 Emergence of additive manufacturing in dentistry

Additive manufacturing provides digitally guided, layer-wise fabrication of dental components for increased customization, reproducibility, and design versatility. The initial research and application of AM in dentistry was in the domain of diagnostic models and surgical guides, but advances in printer resolution, material chemistry, and digital design tools have broadened the application of AM to crowns, bridges, dentures, orthodontic appliances, and implant components [11,12]. Incorporating AM with intraoral scanners, CAD/CAM platforms, and virtual articulators has helped fully digital dental workflows, minimizing physical impressions, remakes, and manual adjustments [13,14]. Bibliometric and scoping studies have shown a major rise of dental AM research output in the past two decades, demonstrating increasing technological maturity and clinical confidence for additively manufactured dental devices [15,16] (Figure 11.2).

11.1.4 Sustainability and life-cycle perspective of dental AM

Sustainable additive manufacturing covers many areas besides operational efficiency, including environmental, economic, and social factors of the

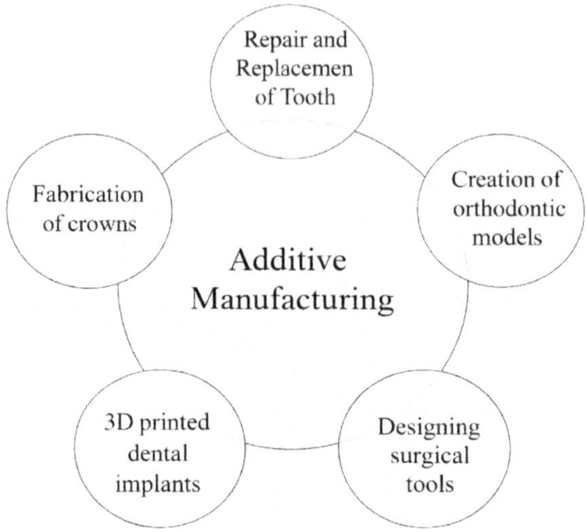

Figure 11.2 Major clinical domains of additive manufacturing in dentistry.

total process life cycle. Life-cycle assessment (LCA) frameworks are being adopted in assessing the sustainability performance of dental AM processes, involving raw material extraction, feedstock preparation, printing, post-processing, clinical use, and end-of-life management [17,18]. AM has been compared to subtractive manufacturing methodologies for the same purpose, often showing higher material utilization efficiency for AM, especially in low-volume, patient-specific dental applications [19]. Nevertheless, the sustainable performance depends more generally upon the choice of process and materials. The energy requirement of polymer AM systems is generally less in comparison to metal and ceramic AM applications, which exhibit higher energy input because of laser operation and high-temperature sintering requirements [20,21]. As a result, sustainability in dental AM is required to be evaluated using holistic life-cycle-based models instead of individual assessment metrics [22] (Figure 11.3).

11.1.5 Materials, properties, and clinical longevity

Material choice is a crucial factor influencing the sustainability performance and clinical durability of dental additive manufacturing. Dental models, splints, and temporary restorations made of polymers such as PEEK, PLA, and PMMA have been widely utilized because of their cost-effectiveness and excellent processability [23,24]. To enhance mechanical characteristics, wear resistance, and service life, researchers are now investigating ways to incorporate fibers and nanoparticles into polymers [25]. Ceramic materials, notably zirconia and glass-ceramics, are also being widely used

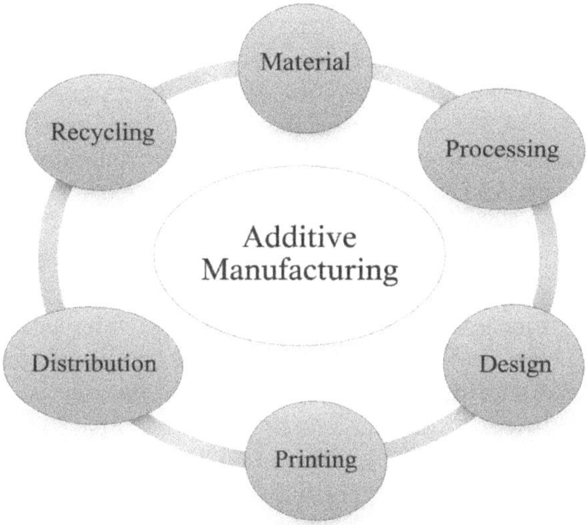

Figure 11.3 Life-cycle perspective of sustainable additive manufacturing in dentistry.

for definitive restorations due to higher esthetics, chemical stability, and their bio-compatibility [26,27]. In general, metal additive manufacturing, especially titanium-based systems, has revolutionized dental implant fabrication through the production of patient-designed porous structures that are improved for osseo-integration and long-term performance [28,29]. Improving clinical longevity in a clinical scope of practice in turn provides a direct correlation to sustainability. Decreased replacement frequency and cumulative resource consumption mean decreased replacement duration leading to decreased usage of resource, which further contributes toward sustainability [30].

11.2 LITERATURE REVIEW

Over the last two decades, the literature on AM has made significant progress. Evolving from exploratory prototyping studies, this research field has grown mature. It now includes advanced manufacturing technologies, material innovation, and digital integration, as well as clinical translation. Bibliometric and scientometric studies repeatedly locate bio-medical and dental applications as one of the fastest-growing research fields in the field of AM [31,32], owing to the growing desire for patient-specific and digitally-enabled healthcare solutions. Initial works dealt with feasibility, dimensional accuracy, and workflow compatibility, while subsequent studies focused on performance optimization, sustainability, and long-term clinical reliability [33]. The expansion of AM-related studies in dentistry is paralleled by the acceptance of digital impression systems, CAD/CAM platforms, and virtual treatment planning solutions. Multiple researches indicate that AM-based workflows contribute to the enhancement of reproducibility, decrease manual intervention, and faster production time compared to traditional subtractive and manual methods [34,35]. In terms of geographic data, much of the research is carried out in North America, Europe and East Asia, and increasingly in emerging countries, AM is proposed as a means for decentralized, affordable, and accessible dentistry engineering, mainly in places that have less laboratory infrastructure [36,37]. A large part of the research is centered on comparative measurements of AM-based technologies used in dentistry in terms of vat photopolymerization, material extrusion, powder bed fusion, hybrid additive–subtractive, and other methods. Vat photopolymerization processes have been observed to provide better geometric characteristics and surface quality and offer superior aesthetic quality, as applicable to crowns, surgical guides, orthodontic appliance applications [38,39]. Extrusion systems have been extensively examined for an economy, as well as suitability for the thermoplastic phase [40], despite problems with surface roughness and anisotropic mechanical properties. Powder-based processes, including selective laser sintering and laser powder bed fusion, have been systematically studied in metal or ceramic dental materials, because of their high strength, complex construction, but the large energy required and post-treatment complexity, however, is one of the enduring problems [41,42]. Material-based research is the secondary paradigm in the literature. Polymeric material (polymethyl methacrylate, polylactic acid (PLA), polyetheretherketone, etc), being highly processed and suitable for provisional restorations, splints, and removable prosthesis, has also been widely studied [43,44]. More recent investigations focused on reinforcement methods by such as the use of fibers, nanoparticles, and a hybrid filler to achieve mechanical strength, wear resistance, and surface finish [45]. In definitive restorations, ceramic additive manufacturing systems such as zirconia- and glass-ceramic-based systems have attracted much interest, with some previous studies reported on slurry-based printing and robocasting improvements, some other problems such as shrinkage, defect development, and sintering-enhanced distortion [46,47]. Metal AM literature, which focus mainly on titanium and its alloys, highlights the clinical importance for porous patient-specific implant design that induces osseo-integration and bio-mechanical compatibility with adjacent bone [48,49]. Recent dental AM studies increasingly focus on sustainability and environmental performance. Comparative studies have shown that AM generally provides a higher material utilization efficiency than subtractive manufacturing, particularly for customized dental components [50]. Nevertheless, several studies warn that significant environmental advantage is process/material and material dependent and metal/ceramic AM systems typically have high energy demand for laser operation and high-temperature sintering at high resolution [51]. LCA–based approaches show that polymer-based AM systems seem to have smaller energy consumption and carbon footprint compared to metal AM systems, but underline the necessity of considering the post-processing, the usage of materials, and also the use of the appropriate equipment and material sourcing [52,53]. Moreover, some research focuses on indirect sustainability benefits of the digital dentistry workflow (less remakes, improved logistics, and reduced clinical chair time) [54]. The association of additive manufacturing and digital intelligence is an increasingly emerging area in research. There is also evidence that the parameter optimization on a dimensional model and its use on the surface and by AI will lead to a material waste and a longer clinical life for additive dental parts [55,56]. Such in-person investigations have validated enhancements in workflow efficiency and treatment predictability that occur between prosthodontics, orthodontics, and implantology [57]. Despite the extensive body of recent research, there are specific gaps including an insufficient number of system-level sustainability assessments, limited long-term clinical performance data on definitive restorations, and lack of perspectives from low- and middle-income countries [58,59]. The deficiencies of these paradigms have motivated recent integrative reviews that aim to develop holistic frames that align materials science, manufacturing technology, sustainability, and clinical dentistry [60–70] (Tables 11.1 and 11.2).

Table 11.1 Technology–material–sustainability mapping of dental additive manufacturing

References	*Focus area*	*Manufacturing technology*	*Material system*	*Sustainability dimension*
[31,32]	Global research trends	Multi-technology	Polymer/ceramic/Metal	Research growth, regional sustainability
[33–35]	Digital dentistry evolution	CAD/CAM+AM integration	Polymeric & hybrid	Workflow efficiency, waste minimization
[36,37]	Emerging economy adoption	Desktop/low-cost AM	Thermoplastics	Decentralized manufacturing, accessibility
[38,39]	Precision dental fabrication	SLA/DLP	Photopolymer resins	Remake reduction, material efficiency
[40]	Cost-driven AM	FDM	Thermoplastic polymers	Energy efficiency, affordability
[41,42]	High-strength components	SLS/LPBF	Metals & ceramics	Energy intensity, powder reuse
[43,44]	Polymeric dental materials	SLA/FDM	PMMA, PLA, PEEK	Recyclability, low embodied energy
[45]	Reinforced materials	Multi-process	Polymer composites	Durability-based sustainability
[46,47]	Ceramic AM	DLP/robocasting	Zirconia, glass-ceramics	Sintering energy, defect sensitivity

(Continued)

Table 11.1 (*Continued*) Technology–material–sustainability mapping of dental additive manufacturing

References	*Focus area*	*Manufacturing technology*	*Material system*	*Sustainability dimension*
[48,49]	Implant manufacturing	LPBF	Titanium alloys	Longevity-driven sustainability
[50,51]	Manufacturing comparison	AM vs subtractive	Mixed	Material savings vs energy trade-off
[52,53]	Life-cycle assessment	Process-level LCA	Polymer/metal	Carbon footprint, energy modeling
[54]	Digital workflow impact	End-to-end digital	Mixed	Logistics reduction, remake avoidance
[55,56]	Intelligent AM systems	AI-assisted AM	Mixed	Process optimization, waste minimization

Table 11.2 Clinical relevance, research gaps, and knowledge contribution in dental AM

References	*Clinical domain*	*Identified research gap*	*Key knowledge contribution*
[31,32]	General dentistry	Limited clinical linkage	Establish global maturity of dental AM research
[33–35]	Digital dentistry workflows	Sustainability metrics underreported	Demonstrate efficiency gains via digital integration
[36,37]	Public & community dentistry	Sparse LMIC-focused studies	Highlight AM for accessible dental care
[38,39]	Prosthodontics, guides	Long-term performance data	Validate AM for accuracy-critical applications
[40]	Provisional restorations	Surface quality limitations	Assess cost–quality trade-offs
[41,42]	Implants, frameworks	Energy-intensive fabrication	Identify performance–energy conflict
[43,44]	Removable prostheses	Aging and durability data	Establish polymers for provisional use
[45]	Load-bearing restorations	Clinical translation limited	Extend service life via reinforcement
[46,47]	Definitive ceramic restorations	Shrinkage and distortion control	Define barriers in ceramic AM
[48,49]	Dental implants	Long-term clinical validation	Link porous AM implants to osseo-integration
[50,51]	Cross-domain dentistry	Overgeneralized sustainability claims	Emphasize context-specific assessment
[52,53]	Sustainable dentistry	Standardized LCA lacking	Introduce LCA as evaluation benchmark

(*Continued*)

Table 11.2 (*Continued*) Clinical relevance, research gaps, and knowledge contribution in dental AM

References	*Clinical domain*	*Identified research gap*	*Key knowledge contribution*
[54]	Chairside dentistry	Quantification of indirect benefits	Highlight systemic sustainability gains
[55,56]	Smart dental labs	Clinical adoption barriers	Demonstrate AI-enabled quality control
[57–59]	Multi-specialty dentistry	Fragmented sustainability evaluation	Expose system-level research gaps
[60–70]	Future dental manufacturing	Lack of unified frameworks	Call for integrative circular-economy models

11.3 METHODOLOGY

Based on this, the review utilizes a structured, transparent, and reproducible approach to synthesize global knowledge on sustainable additive manufacturing in dentistry. The topic is inter-disciplinary, covering manufacturing technologies, materials science, environmental sustainability, clinical dentistry, etc., so a mixed review methodology including narrative synthesis, bibliometric interpretation, and sustainability-oriented analysis was used [71]. It was established based on principles and guidelines set out earlier in the chapter for assessing high-impact review articles, such as in this one [72]. The process of literature identification was limited to peer-reviewed journal articles, conference proceedings, as well as to relevant review papers in internationally renowned databases. Focus was given toward updated and high influence publications related to additive manufacturing technologies, dental application, material parameters, sustainability criteria, life-cycle assessment, and digital dentistry incorporation [73]. Record duplicates were eliminated and screening of relevance from titles, abstracts, and full text articles was performed to ascertain the compatibility with sustainability of additive manufacturing in dentistry [74]. The selected literature was organized into thematic domains, including manufacturing technologies, material systems, sustainability and life-cycle analysis, digital embedding, and clinical applications, for more analytical depth. Thematic analysis allowed for an across-discipline comparison of additive and conventional manufacturing paradigms, as well as the identification of convergence trends and research gaps [75]. Using bibliometric markers such as publication growth, theme clustering, and geographical research contributions, the field's evolution and maturity were contextualized [76]. This review's methodology was mostly based on the sustainability evaluation. Using life-cycle thinking, the data on the material used for additively-made dental equipment, energy consumption, post-processing requirements, and clinical service life were analyzed [77]. The review followed the present sustainable manufacturing paradigms in protecting the environment by weighing the pros and cons of environmental effect, mechanical performance, and clinical durability rather than sustainability as a standalone metric [78]. In this way, it was possible to compare dental AM systems made of polymeric, ceramic, and metallic materials in a meaningful way [79]. Additive manufacturing technology and material reported in literature have been compared based on comparative analysis. Some of the important comparison criteria were dimensional accuracy, mechanical behavior, surface integrity, bio-compatibility, energy demand, and waste generation [80]. Qualitative data from experimental, in vitro and clinical studies were synthesized to determine performance patterns and limitations to avoid bias due to too many trials in areas without standard testing protocols [81]. To support methodological rigor this review adhered to basic systematic and narrative reviews in the engineering and bio-research community. This review was based on systematic inclusion criteria, thematic mapping, and synthesis strategy, and made clear representations of the method used for inclusion to improve reproducibility and credibility [82]. The integration of literature with sustainability literature and with clinical-focused studies was intentionally highlighted, as this was clearly recognised to often unmet need among existing reviews [83]. Finally, the framework of methodology was developed to facilitate prospective, forward-looking exploration. Emerging trends related to AI-assisted manufacturing, circular economy incorporation, and net-zero healthcare manufacturing were included toward the synthesis to keep it contemporary to future research and policy directions [84,85]. This overall approach enables the robustness of the conclusions and promotes the adoption of a global, sustainable-focused view of additive manufacturing in dentistry.

11.4 SUSTAINABILITY AND LIFE-CYCLE ASSESSMENT

Sustainability has become a major evaluative component in additive manufacturing (AM) research nowadays, especially in healthcare and dentistry industries where decisions related to manufacturing have a direct influence on environmental performance, financial viability, and patient outcomes. Recent studies indicate that sustainability of dental AM cannot just be determined in terms of material savings or geometric efficiency and will require holistic life-cycle assessment (LCA) frameworks enabling a measurement of cumulative environmental burdens over the complete product life cycle [86,87]. These frameworks are in line with a wider sustainable production agenda that is guided by combined economic, environmental and social performance measures. Life-cycle assessment criteria on AM are usually concerned with stages: the extraction of raw materials, feedstock development, manufacture and printing, post-processing, clinical application, and end-of-life disposal or recycling [88]. From dentistry perspective, such evaluations show that additive processing is proven to be a more efficient means for material treatment than subtractive milling, especially for tooth restoration and appliances that are specific to the patient [89]. However, several researches advise that even less material waste does not automatically lead to lower environmental concern, because a higher energy usage and post-processing demands would cancel out the material saving effect [90]. Energy consumption is a significant factor contributing to the environmental burden on dental AM technology. Polymer AM technologies such as vat photopolymerization and material extrusion systems often have a comparative advantage concerning energy efficiency on the per part basis [91]. On the other hand, powder-based metal and ceramic AM processes, particularly laser powder bed fusion and ceramic photopolymerization, are more demanding on energy requirement through laser operation, thermal debinding, and sintering stages [92,93]. Life-cycle studies have shown that these heavy energy steps lead to the vast majority of global warming potential, and carbon footprint measurements [94]. Material selection also impacts sustainability consequences of dental AM. Polymers (PMMA and PLA) also have claimed to use less embodied energy and are claimed to have greater potential for recycling/downcycling, particularly than ceramics and metals [95]. Emerging studies of applications in the circular economy emphasize polymer recycling, powder reusing and closed-loop material solutions as potential to minimize dental manufacturing environmental impact [96,97]. Despite this, recycling dental materials in large volumes, on a large scale, is constrained by degraded materials, chemical contamination and regulatory issues, which are a serious obstacle to the development of sustainable dental materials as well as the widespread use of dental devices [98]. In addition to manufacturing-side effects on these findings, emerging research highlights the importance of indirect and system-level sustainability benefits of digital dentistry workflows. Studies show that AM-enabled digital workflows can help reduce remakes in clinics, reduce impression materials, reduce the clinical chair time and optimize logistics as they assist decentralized manufacturing models, for example, with their reduction to cost [99,100]. These indirect benefits are more critical in dentistry, where repetitive modifications and remanufacturing are the leading causes in cumulative consumption of physical and human resources [101]. Many studies show that AM can contribute to overall lowering of production costs of customized dental devices with regards to labor intensity, tooling input, and inventory [102]. Decentralized AM adoption has been identified as a mechanism for gaining economic resiliency and local manufacturing capacity, especially in developing countries, by small and medium-sized dental laboratories and clinics [103,104]. Nonetheless, the initial equipment and training costs in the initial implementation of such systems are challenges [105]. Social sustainability aspects including accessibility, patient-centered care or occupational health have been discussed increasingly in recent research papers. AM-driven customization supports minimally invasive treatment strategies and patients' best outcomes, in accordance to preventive and personalized dentistry paradigms [106]. Moreover, optimized AM post-processing protocols lead to an increasingly safer work environment in dental labs; there are fewer hazardous materials or chemicals for use as much or more than ever before. Despite increasing research effort, literature also indicates that there exists, yet no standardized and harmonized sustainability metrics nor harmonized LCA methodologies specific to dental AM. Numerous studies report patchy system boundaries, functional units, and impact categories and fail to establish or compare cross-comparisons between technologies and materials [107,108]. Contemporary reviews call for the development of sustainability assessment frameworks specialized in the field of dentistry, including mechanical performance, clinical longevity, and environmental impact as part of a unified decision-making tool [109–111]. New directions are increasingly expanding the dialogue in sustainability to intelligent and autonomous manufacturing as well. AI based process control, in-situ monitoring and predictive maintenance has been reported to decrease the defect rates; of course, material waste and energy consumption for AM systems [112,113]. Combined with circular economy approaches and Industry 5.0 principles, these strategies provide a pathway to a net zero and climate-resilient dental manufacturing ecosystem in combination [114–119] (Table 11.3).

Table 11.3 Sustainability and life cycle assessment

Sustainability dimension	*Key focus*	*Representative findings*
Life-cycle assessment	System boundaries, functional units	Need for holistic, standardized LCA
Energy consumption	Process-dependent energy demand	High energy in metal/ ceramic AM
Material efficiency	Waste reduction, near-net shaping	AM superior to subtractive methods
Circular economy	Recycling, powder reuse	Polymer recycling promising but limited
Digital workflow impact	Logistics, remakes, chair time	Significant indirect sustainability gains
Economic sustainability	Cost, decentralization	Benefits for SMEs and local labs
Social sustainability	Patient care, safety	Supports personalized, safer dentistry
Methodological gaps	Metrics, comparability	Lack of dentistry-specific frameworks
Intelligent manufacturing	AI, monitoring, optimization	Enables waste and energy reduction

11.5 ADDITIVE MANUFACTURING TECHNOLOGIES IN DENTISTRY

Additive manufacturing (AM) technologies have transformed the field of dentistry by facilitating the patient-specific fabrication of complex shapes by digital means with careful material management. Vat photopolymerization, material extrusion, powder bed fusion, and emerging hybrid processes have been consistently reported in the literature as the preferred AM technologies utilized in dentistry, with varying degrees of accuracy, material compatibility, productivity, and sustainability [120–122]. Stereolithography (SLA) and digital light processing (DLP) are among the most adopted methods for vat photopolymerization in dentistry practice. These methods are extensively reported for the synthesis of dental models, surgical guides, splints, crowns, and orthodontic fixtures, because of their strong dimensional precision, fine surface layer, and high degree surface resolution [123,124]. The layer-by-layer photopolymer curing process also provides greater geometric perfection than extrusion-based, and vat photopolymerizing is most suitable for accuracy-critical dental treatments under which [125]. Nonetheless, many studies pointed out the problems with resin toxicity, post-curing conditions, and the weak recyclability of the photopolymers, affecting the quality of environment and also with the occupational safety [126]. Material extrusion-based AM, commonly implemented through fused deposition modeling (FDM), is often designed for use in the dental industry where economical and material flexibility options are attractive. The FDM systems work on a variety of thermoplastic polymers, such as PLA, ABS, PMMA, and PEEK, and have been successfully used in dental system, provisional appliance, pilot appliance production, and also in research designs [127,128]. Although the extrusion technology provides cost-effective and simple to operation options, several limitations exist in the literature such as anisotropic mechanical characteristics, poor surface properties, and low dimensional accuracy compared with vat photopolymerization techniques [129]. These obstacles limit the general application of FDM in clinic to provide true restorations, but establish it as an educational, diagnostic, and low-cost method for fabrication [130]. Powder bed fusion processes such as selective laser sintering (SLS) and laser powder bed fusion (LPBF) are indispensable toward advanced mechanical and ceramic metal and ceramic dental components with a high performance. Metal AM based on LPBF has been well studied for dental implant technology, for the preparation of personalized titanium implants with porosity and surface topology that are tailor-made and controlled to the patients' situation for better osseo-integration [131,132]. Ceramic powder-based and slurry-assisted AM systems have been commonly reported for zirconia dental prostheses, although powder flowability, defect generation, and energy-intensive sintering are many challenges [133,134]. Hybrid and upcoming AM technologies are the newest sector in dental manufacturing research. The hybrid additive–subtractive systems combine AM fabrication processes with the milling or machining processes to enhance dimensional precision and surface integrity [135], especially in the context of precision-critical dental restorations. The multi-material AM approaches provide the possibility to fabricate functionally graded dental components with a strong structural property, in addition to esthetic or bioactive properties [136]. Moreover, volumetric and multi-axis AM have been introduced to lower the build period, reduce supporting structures, and gain better mechanical isotropy, alleviating major weaknesses of conventional layer-based processes [137,138]. A sustainability dimension to environmental performance across AM technologies is highlighted by the literature. Vat photopolymerization and extrusion-based systems are characterized by lower energy per unit than powder-based metal and ceramic AM systems and powder bed fusion methods have better material performance and longevity, which could compensate for increased manufacturing energy with longer clinical service life [139–141]. In turn, the choice of technology in dentistry is a compromise between the accuracy, materials, energy use, and clinical longevity, not in terms of a single optimization criterion [142]. Challenges with tooth AM are still arising however, many of the challenge is post-processing and standardization, equipment cost, and regulatory compliance even for dental AM technology and the development process. Recent reviews underline the importance of harmonized process qualification regulations and standards specific to dentistry, and technology-agnostic sustainability benchmarks in order to advance broader clinical utilization [143,144]. All of these points highlight the relevance of technology-aware and sustainable alternative to additive manufacturing choice in dentistry.

11.6 DISCUSSION

In recent years, literature study describes that the effectiveness of the additive manufacturing approaches for dentistry is the result of the embedded coupling between material systems and their processing-induced characteristics, resulting in a functional performance under clinical conditions. With additive manufacturing, unlike other manufacturing routes, anisotropy, layer-wise bonding, and microstructural heterogeneity emerge in these materials with material behavior very much influenced by fabrication strategy and post-processing protocols [145–147]. Thus, the determination on dental additive manufacturing materials needs to go beyond nominal material properties to performance-based and application-oriented evaluation frameworks. Polymeric materials have always been the most common class of materials in dental additive manufacturing in which they feature easy fabrication, affordability, and compatibility with digital workflows. Photopolymer resins and thermoplastic polymers; PMMA,

PLA, and PEEK are widely reported for temporary restorations, orthodontic devices, and surgical guides [148–151]. However, performance evaluations show that limitations of fatigue resistance, wear resistance, and long-term stability especially for oral practice with thermal cycling, moisture, and chemical challenge have been reported [152–154]. These results show that whereas the polymer systems are technically advanced, their clinical use for durability has been a major challenge for definitive applications. Ceramic materials, particularly zirconia type systems possess high esthetic quality, chemical stability, and a high degree of bio-compatibility, providing strong prospects on definitive restoration. Additive production of ceramics facilitates complex contour and material-efficient production, and yet the literature repeatedly reports shrinkage with debinding and sintering, defect sensitivity, and microstructural variation as big obstacles to repeatability [155–157]. Mechanical analyses have already shown a decreased fracture resistance, due to small defects introduced during printing or thermal treatment, highlighting the necessity for rigorous process control and standard sintering process [158]. For metal systems, which is mainly titanium and its alloys, the best similarity exists between the additive manufacturing ability and the requirement of clinical performance purposes. There is constant evidence that as a part of the additive manufacturing, controlled porosity and surface topography help for osseo-integration and bio-mechanical compatibility for dental implant [159–161]. On the performance front, metal additive manufacturing gives high strength and fatigue resistance, but energy consumption, as well as powder reusing and the need for post-processing in processing systems, are concerning to the sustainability, scalability, and efficiency [162]. These trade-offs emphasize that metallic dental materials should be studied from a performance–service-life standpoint rather than a manufacturing-stage perspective in terms of performance alone. Composite and hybrid systems of materials have been explored more frequently to overcome the limitations of monolithic materials. Reinforced polymers, functionally graded (FG) materials and bioactive coatings that have been prepared by MDA techniques, or combined with organic additives, aim at balancing mechanical robustness and biological activities [163–165]. Performance evaluations suggest that they can alleviate anisotropy, increase wear resistance, and maximize the implant–tissue interface [166], but are limited by the complexity of the processing and lack of uniform validation standards. In all material classes, performance evaluations indicate that mechanical behavior, surface integrity, and biological response is mainly determined by the build orientation, interlayer bonding, and post-processing condition. Comparative analysis of them against their own products is impeded by variations in testing methods in studies and complicates an evidence-based material choice [167–169]. Longer time-of-age and bio-compatibility tests further support the conclusion that the residual monomers, metal-ion release, and surface degradation can be detrimental to clinical safety and performance without proper consideration of processing protocols [170]. One important finding from this literature is the disjointness of research in practice. Frequently materials development, performance testing, and clinical validation are conducted in isolation from each other leading to deficient frameworks for decision-making in dental additive manufacturing. Recent integrative studies highlight the necessity of integrating methods to develop integrated evaluation models which not only consider material composition, processing effects, and mechanical and biological properties but also investigate the long-term clinical stability [171].

11.7 CONCLUSION

The present review relies on the coordinated optimization of materials, processing strategies, and performance validation, not independent progress in technology. Polymeric materials account for most dental additive manufacturing to date, because of their accessibility and digital compatibility, but the durability of polymeric materials to long-term oral operation is a limiting factor toward true application. Ceramic and metallic systems may provide superior mechanical strength, esthetics, and biological performance especially for load-bearing and implant applications, however are limited due to more complexity in processing, energy demand, and sensitivity to manufacturing faults. New strategies such as composite and hybrid materials have emerged as a prospective solution to address these trade-offs when combined with structural behavior and bio-compatibility. Performance evaluation is revealed as an important bottleneck in clinical translation. Variability in testing techniques, a lack of long-term aging data and the incomplete synthesis of biological, mechanical, and surface factors hinder the comparability from cross studies and evidence-based material selection. The literature reviewed reiterates the importance of harmonization or dental-specific standardized measures of performance based on actual oral service conditions and clinical longevity.

From a sustainable point of view, additive manufacture gives several positive aspects such as material economy, custom fit, and digital workflow integration, they are related to the material use, process energy consumption and durability. This implies that the sustainability of dental additive manufacturing should be evaluated by life-cycle and durability-based methods instead of manufacturing-stage measures alone. In the long run, however, future capabilities for dental additive manufacturing are integrated systems to integrate advanced materials, innovative manufacturing, standardized performance evaluation, and life-cycle thinking. Such convergence is vital to realize clinically effective, sustainable, and environmental friendly dental treatments for the world—making sustainable additive manufacturing a fundamental pillar of future-advanced oral health care.

REFERENCES

1. Klenam, D., Asumadu, T., Bodunrin, M., Obiko, J., Genga, R., Maape, S., & Soboyejo, W. (2025). Toward sustainable industrialization in Africa: The potential of additive manufacturing–an overview. *Frontiers in Manufacturing Technology*, *4*, 1410653.
2. Sala, D., & Richert, M. (2025). Perspectives of additive manufacturing in 5.0 industry. *Materials*, *18*(2), 429.
3. Shinkai, R. S. A., Biazevic, M. G. H., Michel-Crosato, E., & de Campos, T. T. (2025). Environmental sustainability related to dental materials and procedures in prosthodontics: A critical review. *The Journal of Prosthetic Dentistry*, *133*(6), 1466–1473.
4. Ben Said, L., Ayadi, B., Alharbi, S., & Dammak, F. (2025). Recent advances in additive manufacturing: A review of current developments and future directions. *Machines*, *13*(9), 813.
5. Ashok, G., Kumar, P., & Prabhu, T. R. (2025). Global developments in additive manufacturing of polymer composite materials: A scientometric review. *Archives of Computational Methods in Engineering*, 1–22. https://doi.org/10.1007/s11831-025-10123-x
6. Gahletia, S., & Garg, R. K. (2025). Dismantling barriers in integrating patient-centred care with additive manufacturing to assess the fit of orthodontic retainers for futuristic preventative healthcare. *Progress in Additive Manufacturing*, *10*(2), 1063–1084.
7. Eliades, T., Adel, S. M., Sercan, A., Atsawasuwan, P., Foong, K. W., Hiskia, A., ... Vaiid, N. (2025). Environmental footprints in orthodontics: The World Federation of Orthodontists' white paper on sustainable practices, challenges and research imperatives. *Journal of the World Federation of Orthodontists*, *14*(4), 194–201.
8. Vasishta, B. V., Preetham, T. H., Akhil, U. V., & Radhika, N. (2025). An overview of the evolution of additive manufacturing from 3D to 5D printing in biomedical applications. *Progress in Additive Manufacturing*, *10*(5), 3287–3303.
9. Aldahash, S. (2025). Implementation of additive manufacturing technologies for the growth and sustainability of manufacturing SMEs in Saudi Arabia: An exploratory study. *Sustainability*, *17*(3), 1249.
10. Subramani, R., Leon, R. R., Nageswaren, R., Rusho, M. A., & Shankar, K. V. (2025). Tribological performance enhancement in FDM and SLA additive manufacturing: Materials, mechanisms, surface engineering, and hybrid strategies—A holistic review. *Lubricants*, *13*(7), 298.
11. Mathur, V., Dsouza, V., Srinivasan, V., & Vasanthan, K. S. (2025). Volumetric additive manufacturing for cell printing: Bridging industry adaptation and regulatory frontiers. *ACS Biomaterials Science & Engineering*, *11*(1), 156–181.
12. Zafar, M. Q., Sajjad, R., Anwar, M. T., Mustafa, M. S., Husnain, N., Aqeel, A. B., & Khan, M. B. (2025). A review on metal additive manufacturing-types, applications and future trends. *Recent Progress in Materials*, *7*(1), 1–24.
13. Rahmani, R., Bashiri, B., Lopes, S. I., Hussain, A., Maurya, H. S., & Vilu, R. (2025). Sustainable additive manufacturing: An overview on life cycle impacts and cost efficiency of laser powder bed fusion. *Journal of Manufacturing and Materials Processing*, *9*(1), 18.

14. Kumar, S., & Kumar, R. (2025). A comprehensive study on Additive Manufacturing techniques, Machine Learning integration, and Internet of Things-driven sustainability opportunities. *Journal of Materials Engineering and Performance*, 1–68. https://doi.org/10.1007/s11665-025-10757-x
15. Al-Masa'fah, W., Abushaikha, I., & Bwaliez, O. M. (2025). Exploring the role of additive manufacturing in the prosthetic supply chain: Qualitative evidence. *The TQM Journal*, *37*(7), 1963–1991.
16. Singh, J., Vasudev, H., Prakash, C., & Singh, M. (Eds.). (2025). *Advanced Manufacturing Technologies in Biomedical Science: Practical Applications, Case Studies, and Future Trends*. Boca Raton, FL: CRC Press.
17. Romario, Y. S., Bhat, C., Jiang, W. R., Chou, I. C., Weng, D. Y., Ramezani, M., & Jiang, C. P. (2025). High-precision low-cost fabrication of dental surgical guides using dental grade photocurable resin through a hybrid additive-subtractive machine with reverse scanning. *Rapid Prototyping Journal*, *31*(5), 1028–1038.
18. Yüceer, Ö. M., Kaynak Öztürk, E., Çiçek, E. S., Aktaş, N., & Bankoğlu Güngör, M. (2025). Three-Dimensional-Printed photopolymer resin materials: A narrative review on their production techniques and applications in dentistry. *Polymers*, *17*(3), 316.
19. Mathur, M., Phogat, P., Jewariya, M., & Wan, M. (2025). Surface characteristics and their influence on osseointegration: A scientometric analysis with a focus on dental implants. *Journal of Maxillofacial and Oral Surgery*, 1–15.
20. Dimitriadis, K., Baciu, D., Koltsakidis, S., Tzetzis, D., Garmpi, E., Roussi, E., ... Andreouli, C. D. (2025). Additive manufacturing of zirconia-based pastes for dental prosthesis via robocasting method. *Journal of Materials Engineering and Performance*, *34*(6), 4735–4749.
21. Dhingra, A. K., Dhingra, S., Singh, A., Rathee, M., Kaushik, H., Kaushik, A., & Chhabra, D. (2025, June). Redefining the supply chain using confluence of artificial intelligence and additive manufacturing. In *2025 4th International Conference on Computational Modelling, Simulation and Optimization (ICCMSO)* (pp. 355–360). Piscataway, NJ: IEEE.
22. Yadav, A., Garg, R. K., Sachdeva, A., Qureshi, K. M., & Qureshi, M. R. N. M. (2025). Strategic framework for additive manufacturing with smart polymer composites: A pathway to net-zero manufacturing. *Polymers*, *17*(10), 1336.
23. Lobo, S., Argolinha, I., Machado, V., Botelho, J., Rua, J., Li, J., & Mendes, J. J. (2025). Advances in digital technologies in dental medicine: Enhancing precision in virtual articulators. *Journal of Clinical Medicine*, *14*(5), 1495.
24. Zarei, A., & Farazin, A. (2025). Synergizing additive manufacturing and machine learning for advanced hydroxyapatite scaffold design in bone regeneration. *Journal of the Australian Ceramic Society*, *61*(3), 797–813.
25. Çakmak, G., Sabatini, G. P., de Paula, M. S., Orgev, A., Kahveci, Ç., Revilla-Léon, M., & Yilmaz, B. (2025). Can nonhazardous postprocessing cleaning solutions enable adequate surface properties for printed dental casts in different resins? *The Journal of Prosthetic Dentistry*, *133*(3), 893–902.
26. Punia, U., & Garg, R. K. (2025). Reinforcing dental restoration material PMMA with SS 316L nanoparticles for mechanical performance analysis. *Progress in Additive Manufacturing*, *10*(4), 3003–3014.

27. Bhat, C., Romario, Y. S., Chou, I. C., Jiang, W. R., Wu, Y. Y., Ramezani, M., & Jiang, C. P. (2025). Development of a fully digital design process for customized mandibular advancement and its precision additive manufacturing. *Annals of 3D Printed Medicine*, *18*, 100199.
28. Miao, W. J., Wang, S. Q., Wang, Z. H., Wu, F. B., Zhang, Y. Z., Ouyang, J. H., ... Zou, Y. C. (2025). Additive manufacturing of advanced structural ceramics for tribological applications: Principles, techniques, microstructure and properties. *Lubricants*, *13*(3), 112.
29. Panahi, O. (2025). Innovative biomaterials for sustainable medical implants: A circular economy approach. *European Journal of Innovative Studies and Sustainability*, *1*(2), 20–29.
30. Hsu, M. H., & Chang, Y. C. (2025). The literature on digital technologies for removable dental prostheses: A two-decade bibliometric analysis. *Journal of Dental Sciences*, *20*(1), 170–177.
31. Cheraghiyan, M. (2025). Nanotechnology in dentistry: Potential applications and future perspectives. *Journal of Oral and Dental Health Nexus*, *2*(1), 1–13.
32. de Paula Netto, M. O., Carvalho, C. S., Rodriguez, T. T., de Stefano, E., Júnior, M. L. L., & Castañõn, J. A. B. The use of additive manufacturing in the development of new products and technologies in brazil: A systematic review. *Journal of Manufacturing Technology Management*, *35*(2), 310–335.
33. Estévez, R., Quijada-Maldonado, E., Romero, J., & Abejón, R. (2025). Additive manufacturing and chemical engineering: Looking for synergies from a bibliometric study. *Applied Sciences*, *15*(6), 2962.
34. Li, C., Wang, G., Wang, S., Shen, W., Zhang, Y., Kang, J., & Xiao, Z. (2025). Fluorapatite glass-ceramics in dentistry: Synthesis, properties, forming technology, applications, challenges, and future perspectives. *Materials*, *18*(4), 804.
35. Lepišová, M., Tomášik, J., Oravcová, Ľ., & Thurzo, A. (2025). Three-dimensional-printed elements based on polymer and composite materials in dentistry: A narrative review. *Bratislava Medical Journal*, *126*(1), 14–27.
36. Brescia, V., Cavarra, F., Campra, M., Boffano, P., & Rocchetti, V. (2025). Bibliometric analysis and trends related to dental tourism: Qualitative, social, economic and ethical implications. *Health Services Management Research*, 09514848251314111. https://doi.org/10.1177/09514848251314111
37. Cioloca Holban, C., Tatarciuc, M., Vitalariu, A. M., Vasluianu, R. I., Antohe, M., Diaconu, D. A., ... Dima, A. M. (2025). Three-dimensional printing and CAD/CAM milling in prosthodontics: A scoping review of key metrics towards future perspectives. *Journal of Clinical Medicine*, *14*(14), 4837.
38. Sherin, T., Motapothula, M. R., Dalapati, G. K., Ramakrishna, S., Sangaraju, S., Chakrabortty, S., ... Ghosh, S. (2025). A comprehensive review on realization of self-cleaning surfaces by additive manufacturing. *Progress in Additive Manufacturing*, *10*(4), 1785–1808.
39. Patadiya, J., Kandasubramanian, B., Sreeram, S., Patil, P. D., Mujawar, R., Indalkar, A., ... Aldawood, F. K. (2025). Strategic implementation of multi-material additive manufacturing: Bridging research and real-world applications. *ACS Omega*, *10*(14), 13749–13762.

40. Sharma, A., & Bharti, P. S. (2025). Optimization of resin printing parameters for improved surface roughness using metaheuristic algorithms: A multifaceted approach. *Journal of Materials Engineering and Performance*, *34*(10), 8525–8539.
41. Haiati, S., Dotchev, K., & Lowther, M. (2025). Utilizing powder bed fusion additive manufacturing technology to fabricate parts with controlled porosity and permeability characteristics for filtration applications. *International Journal of Precision Engineering and Manufacturing-Green Technology*, *12*(1), 135–149.
42. Ramos, A., Angel, V. G., Siqueiros, M., Sahagun, T., Gonzalez, L., & Ballesteros, R. (2025). Reviewing additive manufacturing techniques: Material trends and weight optimization possibilities through innovative printing patterns. *Materials*, *18*(6), 1377.
43. Mosadegh, M., Khakzad, M., Sepasi, Z., Nandigama, K., Kumar, G., & Minary-Jolandan, M. (2025). Single-step thermal debinding for ceramics vat photopolymerization in less than 30-minutes. *Ceramics International*, *51* (21, Part B), 34846–34857. https://doi.org/10.1016/j.ceramint.2025.05.206
44. Borthakur, P. P., Das, A., Sahariah, J. J., Pramanik, P., Baruah, E., & Pathak, K. (2025). Revolutionizing patient care: 3D printing for customized medical devices and therapeutics. *Biomedical Materials & Devices*, *3*(2), 1–28. https://doi.org/10.1007/s44174-025-00324-2
45. Danila, A. I., Breban-Schwarzkopf, D., Daescu, E., Olariu, I., & Dinu, S. (2025). Systematic review of the quality of stereolithographic three-dimensionally printed materials for provisional dental restorations. *Materials*, *18*(3), 721.
46. Abbasi, M., Váz, P., Silva, J., & Martins, P. (2025). Head-to-Head evaluation of FDM and SLA in additive manufacturing: Performance, cost, and environmental perspectives. *Applied Sciences*, *15*(4), 2245.
47. Bhardwaj, D., Siraj, I., Singh, A., Deshwal, D., Kaushik, A., Bhiogade, G., & Gahletia, S. (2025). Addressing critical challenges encountered during extrusion-based ASA additive manufacturing: A process optimization approach. *Journal of Materials Engineering and Performance*, *35*, 1–16. https://doi.org/10.1007/s11665-025-11996-8
48. Sharma, R., Verma, N., Dewangan, R., Sharma, A. K., Tripathi, M., & Pani, B. (2025). Delineating the techniques employed for the fabrication of sustainable polymers via 3D printing phenomena. In *Sustainable 3D Printing for Innovative Biopolymer Production and Applications* (pp. 25–47). Boca Raton, FL: CRC Press.
49. Alharbi, N. M., & Osman, R. B. (2025). Trueness of multichromatic versus monochromatic veneers printed using tilted stereolithography technology at different build angles? An in vitro study. *Journal of Dentistry*, *156*, 105702.
50. El-Bassyouni, G. T., Mouneir, S. M., & El-Shamy, A. M. (2025). Advances in surface modifications of titanium and its alloys: Implications for biomedical and pharmaceutical applications. *Multiscale and Multidisciplinary Modeling, Experiments and Design*, *8*(5), 265.
51. Khan, S. B., Irfan, S., Zhang, Z., & Yuan, W. (2025). Redefining medical applications with safe and sustainable 3D printing. *ACS Applied Bio Materials*, *8*(8), 6470–6525.

52. Farrag, A., Yang, Y., Cao, N., Won, D., & Jin, Y. (2025). Physics-informed machine learning for metal additive manufacturing. *Progress in Additive Manufacturing*, *10*(1), 171–185.
53. Bhardwaj, A. (2025). 3D printing and additive manufacturing: Recent trends in fashion and textile production. In *Use of Digital and Advanced Technologies in the Fashion Supply Chain* (pp. 223–246). Boca Raton, FL: CRC Press (Taylor & Francis Group).
54. Domysche, M., Terekhov, S., Astapenko, O., Vefelev, S., & Tatarina, O. (2025). The use of 3D printers for the manufacture of complex multi-component dentures in dentistry. *Salud, Ciencia y Tecnología-Serie de Conferencias*, *4*, 668.
55. Filimonova, E. A., Lozovaya, A. V., Prosyankin, E. E., Mustafina, A. R., & Chapala, P. (2025). Thermal treatment influence on optical properties of 3D printed objects by vat photopolymerization. *Progress in Additive Manufacturing*, *10*(4), 2703–2713.
56. Suva, A. I., & Rahman, K. A. (2025, April). Additive manufacturing and industry 4.0 integration: An approach for Bangladesh's industrial advancement. In *AIP Conference Proceedings* (Vol. 3307, No. 1, p. 020021). *Proceedings of the 2nd International Conference on Mechanical, Industrial and Energy Engineering* (ICMIEE 2024). Melville, NY: AIP Publishing. https://doi.org/10.1063/5.0267944
57. Kumar, A., & Rajak, S. (2025). Selection of bio-implant materials based on metal additive manufacturing using integrated SWARA and WASPAS MCDM technique. *Journal of Materials Engineering and Performance*, *34*(17), 18766–18786.
58. Hussien, Z. Y., Moften, A. Q., & Abdulrehman, M. A. (2025). Review of zirconia (ZrO_2) biomedical applications: Advanced manufacturing techniques and materials properties. *Revue des Composites et des Materiaux Avances*, *35*(2), 345.
59. Rumapea, E. S., Ginting, R., & Ishak, A. (2025). Integration of reverse engineering and additive manufacturing methods in the design and development of prosthetic products: A literature review. *Jurnal Sistem Teknik Industri*, *27*(1), 8–13.
60. Daniyan, I. A., Muvunzi, R., Fameso, F., Ndambuki, J., Kupolati, W., & Snyman, J. (2025). Application of multi-criteria decision and simulation approaches to selection of additive manufacturing technology for aerospace application. *Computers, Materials & Continua*, *83*(2), 1623–1648.
61. Dai, Y., & Lu, Y. (2025). 3D printing driven business model innovation and supply chain operations: A perspective of strategic alignment. *Journal of Manufacturing Technology Management*, *36*(4), 820–839.
62. Vanerio, D., Guagliano, M., & Bagherifard, S. (2025). Emerging trends in large format additive manufacturing processes and hybrid techniques. *Progress in Additive Manufacturing*, *10*(4), 1945–1972.
63. Park, J. Y., Jung, Y. N., Jang, K. J., Lee, S. K., Choi, S. W., Lee, Y. S., ... Yun, K. D. (2025). Effect of axis change on shrinkage rate of 3D-printed bioceramic zirconia fabricated via digital light processing. *Biomimetics*, *10*(3), 140.
64. Zhang, Z., Wang, Y., & Wang, W. (2025). Machine learning in gel-based additive manufacturing: From material design to process optimization. *Gels*, *11*(8), 582.

65. Daghrery, A., Vinothkumar, T. S., Majrashi, H., Faqihi, G., Gofshi, R., Almasoudi, S., ... Chaurasia, A. (2025). Dimensional accuracy of additive and subtractive manufactured ceramic-reinforced hybrid composite inlays: A CBCT-based in vitro study. *Scientific Reports*, *15*(1), 6048.
66. Scheithauer, U., Johne, R., Gottlieb, L., Weingarten, S., Wiemer, H., & Holtzhausen, S. (2025). Recent developments in the field of additive manufacturing of ceramic-based multi-material components open the door to highly functionalized components. *Progress in Additive Manufacturing*, *10*(6), 4051–40591–9.
67. Punia, R., & Kandasubramanian, B. (2025). Towards circular economy: Integrating polymer recycling and 3D printing for economic regeneration. *Polytechnica*, *8*(1), 2.
68. Gustafsson, C., Chirumalla, K., Bruch, J., & Sannö, A. (2025). Additive manufacturing technology integration: An operational capability perspective. *Journal of Manufacturing Technology Management*, *36*(9), 134–153.
69. Lin, W. S., & Chen, L. (2025). 3D-printed zirconia and lithium disilicate in dentistry and their clinical applications. *International Journal of Prosthodontics*, *38*(1), 12–26.
70. Gupta, B., Gupta, S., Bedia, A. S., Aljabri, M. Y., & Elamin, N. M. H. (2025). Role of digital dentistry in prevention and diagnosis - A systematic review. *Bangladesh Journal of Medical Science*, *24*, 160–169.
71. Siddiqui, M. M. U. Z., & Tabassum, A. (2025). Condition-based monitoring techniques and algorithms in 3D printing and additive manufacturing: A state-of-the-art review. *Progress in Additive Manufacturing*, *10*(8), 4257–4304.
72. Agrawal, K., & Bhat, A. R. (2025). Advances in 3D printing with eco-friendly materials: A sustainable approach to manufacturing. *RSC Sustainability*, *3*(6), 2582–2604.
73. Quaranta, A., Hui, W. L., & Bhargava, N. (2025). Applying the VRIO framework to clinical and operative workflows in dental practice management: Sustainable competitive advantage in dental business. *Annali di stomatologia*, *16*(1), 31–36.
74. Yıldırım, Ö., Yeşil, Z., & Hatipoğlu, Ö. (2025). Effect of different 3D-printing systems on the flexural strength of provisional fixed dental prostheses: A systematic review and network meta-analysis of in vitro studies. *BMC Oral Health*, *25*(1), 82.
75. Fan, H., Liu, C., Bian, S., Ma, C., Huang, J., Liu, X., ... Li, B. (2025). New era towards autonomous additive manufacturing: A review of recent trends and future perspectives. *International Journal of Extreme Manufacturing*, 7(3), 032006. https://doi.org/10.1088/2631-7990/ada8e4
76. Teschke, M., Grafe, L., Stammkoetter, S., Tenkamp, J., & Walther, F. (2025). Microstructural, electrochemical, and mechanical assessment of additive manufactured titanium grade 23 for dental implants application. *In Vivo*, *39*(3), 1751–1766.
77. Kumar, P., & Singh, J. (2025). 3D printed materials in environmental remediation. In *Nanomaterials in Environmental Remediation*, edited by Nisha Kumari, Anuj Kumar, and Vinod Kumar, 283–307. Boca Raton, FL: CRC Press. https://doi.org/10.1201/9781003488828-12
78. Khan, I., Amin, J., Abas, M., Babar, M., Mikail Shah, S., Ali, A., ... Hira, F. (2025). Extrusion additive manufacturing of particle-reinforced polymer composites: Materials, processes, and applications. *Progress in Additive Manufacturing*, *10*(4), 2981–3026.

79. Altun, F., Bayar, A., Hamzat, A. K., Asmatulu, R., Ali, Z., & Asmatulu, E. (2025). AI-driven innovations in 3D printing: Optimization, automation, and intelligent control. *Journal of Manufacturing and Materials Processing, 9*(10), 329.
80. Guo, F., Flores-Mir, C., Su, N., Glenny, A. M., Clarkson, J., & Hua, F. (2025). Translating research evidence into dental practice. *Translational Dental Research, 1*(2), 100017.
81. Jun, M. K., Kim, J. W., & Ku, H. M. (2025). Three-dimensional printing in dentistry: A scoping review of clinical applications, advantages, and current limitations. *Oral, 5*(2), 24.
82. Li, K., Zhang, M., Zhang, Z., Jin, P., Wang, Y., Yan, W., ... Murr, L. E. (2025). High performance realization of functionally graded materials based on integrated optimal design and additive manufacturing: A review. *International Materials Reviews, 70*(6), 497–547.
83. Wang, X., Li, H., Mu, M., Ye, R., Zhou, L., & Guo, G. (2025). Recent development and advances on polysaccharide composite scaffolds for dental and dentoalveolar tissue regeneration. *Polymer Reviews, 65*(1), 47–103.
84. Sharma, B., & Shekhar, S. (2025). The framework of the breakthroughs in the 3D printing technique. In *Sustainable 3D Printing for Innovative Biopolymer Production and Applications*, edited by Bhasha Sharma, Balaram Pani, Shashank Shekhar, and Jude A. Okolie, 1–24. Beverly, Massachusetts, USA: Wiley-Scrivener (John Wiley & Sons). ISBN 9781119791713.
85. Maturi, M., Locatelli, E., de Leon, A. S., Franchini, M. C., & Molina, S. I. (2025). Sustainable approaches in vat photopolymerization: Advancements, limitations, and future opportunities. *Green Chemistry, 27*(29), 8710–8754.
86. Verma, V. K., Kamble, S. S., Ganapathy, L., Mani, V., Belhadi, A., & Shi, Y. (2025). Exploring the barriers in medical additive manufacturing from an emerging economy. *Production Planning & Control, 36*(4), 464–484.
87. Nsanzumuhire, C., Daramola, O. O., Oladele, I. O., & Akinwekomi, A. D. (2025). Review of current progress on additive manufacturing of medical implants and natural/synthetic fibre reinforced composites. *Materialwissenschaft und Werkstofftechnik, 56*(1), 17–42.
88. Vijayan, A., Vishnu, J., Shankar, B., & Sambhudevan, S. (2025). A review on hydroxyapatite fabrication: From powders to additive manufactured scaffolds. *Biomaterials Science, 13*(4), 913–945.
89. García-Hernández, C., García-Cabezón, C., González-Diez, F., Ampudia, M., Juanes-Gusano, D., Rodriguez-Cabello, J. C., & Martín-Pedrosa, F. (2025). Effect of processing on microstructure, mechanical properties, corrosion and biocompatibility of additive manufacturing Ti-6Al-4V orthopaedic implants. *Scientific Reports, 15*(1), 14087.
90. Solouki, A., Abbaslou, M., Aliha, M. R. M., & Bachari, M. S. (2025). Analyzing the impact of hole radii on flexural strength of notched 3D printed components using Machine learning. *Engineering Failure Analysis, 173*, 109401.
91. Szczęsny, G., Kopec, M., & Kowalewski, Z. L. (2025). Toxicity, irritation, and allergy of metal implants: Historical perspective and modern solutions. *Coatings, 15*(3), 361.
92. Oguntuyi, S. D., Nyembwe, K., Shongwe, M. B., Kabasele, J., & Mojisola, T. (2025). A review of the influence of sand properties on parts manufactured by rapid sand casting through additive manufacturing. *The International Journal of Advanced Manufacturing Technology, 136*(5), 1989–2002.

93. Eid, N. K. (2025). A review on the power of CAD/CAM technology and the material science in modern manufacturing. *ERU Research Journal*, *4*(1), 2223–2250.
94. Derakhshani, A. J., Nicic, R., Peroz, I., Peroz, S., Beuer, F., & Prause, E. (2025). Three-dimensional-printed restorations for prosthetic rehabilitation of localized tooth wear using the Dahl concept: A case report. *Journal of Prosthodontic Research*, *69*(3):454–459, https://doi.org/10.2186/jpr.JPR_D_24_00158.
95. Cheadle, A. M., Maier, E., Palin, W. M., Tomson, P. L., Poologasundarampillai, G., & Hadis, M. A. (2025). The impact of modifying 3D printing parameters on mechanical strength and physical properties in vat photopolymerisation. *Scientific Reports*, *15*(1), 12592.
96. Puchakayla, P. K. R., Pegu, B., Gandhi, P., & Singh, G. (2025). Defects evolution during printing, debinding and sintering for additive manufacturing of yttria stabilized zirconia. *Materials Characterization*, *221*, 114752.
97. Alshamrani, A., Fahimipour, A., Alsarani, M., & Ellakwa, A. (2025). Flexural properties and finite element analysis of novel 3D-printed ceramic materials via 3D-LCM technology. *Australian Dental Journal*, *70*(1), 58–69.
98. Puttaswamy, J. T., Puttegowda, M., Yashas Gowda, Y. G., & Vedavathi, D. H. (2025). Evolution and recent advancements of composite materials in rapid prototyping. In *Applications of Composite Materials in Engineering*, edited by Madhu Puttegowda, Yashas Gowda T. G., Binoj J. S., Sanjay Mavinkere Rangappa, and Suchart Siengchin, 169–193. Duxford, United Kingdom: Woodhead Publishing (Elsevier). https://doi.org/10.1016/B978-0-443-13989-5.00007-3
99. Dawood, R. M., & Mahdee, A. F. (2025). Fabrication and characterization of 3D-printed polymeric-based scaffold coated with bioceramic and naringin for a potential use in dental pulp regeneration (in vitro study). *International Endodontic Journal*, *58*(4), 627–642.
100. Kumar, S., Nasna, P., & Ghosh, G. (2025). Recent advancement in biocompatible materials, hybrid bioactive coating, surface modification and post-processing techniques for the fabrication of biomedical implant: Critical review and future prospects. *Proceedings of the Institution of Mechanical Engineers, Part C: Journal of Mechanical Engineering Science*, *239*(9), 3247–3286.
101. Zhang, Y., Zhang, W., Yang, M., Li, M., Zhou, L., Liu, Y., ... Zheng, Y. (2025). Comprehensive review of polyetheretherketone use in dentistry. *Journal of Prosthodontic Research*, *69*(2), 215–232.
102. Du, K., Muao, T., Plaksina, P., Hung, S. T. H., Cameron, A. B., & Choi, J. J. E. (2025). Masking ability of subtractively and additively manufactured dental ceramic restorations: A systematic review. *Journal of Esthetic and Restorative Dentistry*, *37*(6), 1350–1362.
103. Farhadi Nia, M., Ahmadi, M., & Irankhah, E. (2025). Transforming dental diagnostics with artificial intelligence: Advanced integration of ChatGPT and large language models for patient care. *Frontiers in Dental Medicine*, *5*, 1456208.
104. Wakjira, Y., Cioni, A., & Lemu, H. G. (2025). Current status of the application of additive-manufactured TPMS structure in bone tissue engineering. *Progress in Additive Manufacturing*, *10*(2), 1085–1102.

105. Miranda Cid, A., Pérez Mendoza, G. J., Corro Valdez, N., López Perrusquia, N., Doñu Ruiz, M. A., González Rebattú y González, A., & Hernández Acosta, H. Y. (2025). Study of mechanical properties of heat-treated peek for medical devices through additive manufacturing. *Progress in Additive Manufacturing*, *10*(5), 3129–3144.
106. Bhat, C., Jiang, C. P., Shan Romario, Y., Paral, S. K., & Toyserkani, E. (2025). Critical review of metal-ceramic composites fabricated through additive manufacturing for extreme condition applications. *Mechanics of Advanced Materials and Structures*, *32*(10), 2153–2180.
107. Chattopadhyay, J., Srivastava, N., & Pathak, T. S. (2025). Comprehensive review of 3D printing techniques emphasizing thermal characterization in biomedical prototyping. *Journal of Biomaterials Applications*, *39*(9), 971–995.
108. Saqib, M., Ren, W., Arshad, Q., Asghar, A., Raza, M., Hamza, M., & Chen, Z. (2025). Systematic optimization of silicon nitride slurries for high-precision photopolymerization 3D printing. *Ceramics International. 51*(10), 17342–17355.
109. Solanki, A., Ranganath, M. S., & Singholi, A. K. (2025). Review on advancements in 3D/4D printing for enhancing efficiency, cost-effectiveness, and quality. *International Journal on Interactive Design and Manufacturing (IJIDeM), 19*(5), 3153–3169.
110. Abdulakhatov, J. Q., & Oxunov, B. M. (2025). A review of advancements of artificial intelligence in dentistry. *Web of Medicine: Journal of Medicine, Practice and Nursing*, *3*(3), 118–142.
111. Galbraith, A., Doan, M., Galbraith, T., & Abubakr, N. H. (2025). Evaluation of color stability and marginal integrity in provisional restorations: A study of milling, 3D printing, and conventional fabrication methods. *Dentistry Journal*, *13*(5), 189.
112. Fenta, E. W., Tsegaye, A. A., Abere, A. E., & Tefera, G. T. (2025). Opportunities in flexible manufacturing systems in the near future. *Global Journal of Flexible Systems Management*, *26*(2), 145–165.
113. Łosiewicz, B., Osak, P., Nowińska, D., & Maszybrocka, J. (2025). Developments in dental implant surface modification. *Coatings*, *15*(1), 109.
114. Koppunur, R., Ramakrishna, K., Manmadhachary, A., Kumar, D. K., & Sridhar, V. (2025). Topology optimization and manufacturing of maxillofacial patient specific implant using FEA and AM. *Bioprinting*, *48*, e00412
115. Verma, C., Alameri, A., Barsoum, I., & Alfantazi, A. (2025). Review on corrosion-related aspects of metallic alloys additive manufactured with laser powder bed-fusion (LPBF) technology. *Progress in Additive Manufacturing*, *10*(5), 3195–3223.
116. Madival, H., & Rajiv, A. (2025). A comprehensive review of bioactive glasses: Synthesis, characterization, and applications in regenerative medicine. *Biomedical Materials & Devices*, *3*(1), 45–65.
117. Onyia, T. M., Olarinoye, I. A., & Jimoh, S. A. (2025). Advancements and challenges in 3D scanning: A comprehensive review of engineering applications. *African Journal of Advances in Science and Technology Research*, *18*(1), 191–206.
118. Bhowmik, A., Kumar, R., MK, R., Karun Kumar, Y., Samal, P., Mahapatro, A., ... Santhosh, A. J. (2025). Predictive modeling of the mechanical behavior of 3D-printed polylactic acid/wood composite: Comparison of GEP and ANN methods. *AIP Advances*, *15*(4), 045221.

119. Mandurino, M., Cortili, S., Coccoluto, L., Greco, K., Cantatore, G., Gherlone, E. F., ... Paolone, G. (2025). Mechanical properties of 3D printed vs. subtractively manufactured composite resins for permanent restorations: A systematic review. *Materials*, *18*(5), 985.
120. Vyas, J., Raytthatha, N., Vyas, P., Prajapati, B. G., Uttayarat, P., Singh, S., & Chittasupho, C. (2025). Biomaterial-based additive manufactured composite/scaffolds for tissue engineering and regenerative medicine: A comprehensive review. *Polymers*, *17*(8), 1090.
121. Bunyan, S. F., Shakir, S. M., & Zardawi, F. M. M. (2025). Color stability and roughness of ocular prosthesis between heat-cured acrylic and 3D printed acrylic after artificial weathering. *International Journal of Dentistry*, *2025*(1), 6674943.
122. Farjaminejad, R., Farjaminejad, S., Hasani, M., Garcia-Godoy, F., Sayahpour, B., Marya, A., & Jamilian, A. (2025). The role of tissue engineering in orthodontic and orthognathic treatment: A narrative review. *Oral*, *5*(1), 21.
123. de Jesus Vieira, J. P., da Silva, S. S., de Fátima Souza, I., Pedras, M. B., de Avelar Freitas, B. A., & González Torres, L. A. (2025). Study of the biocompatibility of polymeric resins processed by 3D printing for applications in manufacturing of devices for short term 3D cultures. *Microfluidics and Nanofluidics*, *29*(6), 37.
124. Mohammad, A. H., Njim, E. K., Jweeg, M., & Al-Waily, M. (2025). A three-dimensional printed porous PLA polymer implant: A critical review. *Journal of Engineering and Sustainable Development*, *29*(4), 485–494.
125. Ahmad, K. H., Mohamad, Z., Khan, Z. I., & Habib, M. (2025). Tailoring UV penetration depth in photopolymer nanocomposites: Advancing SLA 3D printing performance with nanofillers. *Polymers*, *17*(1), 97.
126. Işik, U., Demir, H., & Özlü, B. (2025). Multi-objective optimization of process parameters for surface quality and geometric tolerances of AlSi10Mg samples produced by additive manufacturing method using taguchi-based gray relational analysis. *Arabian Journal for Science and Engineering*, *50*(12), 9211–9229.
127. Gao, M., Xu, J., Liu, K., Zhang, S., & Tan, J. (2025). Material and energy co-optimization for additive manufacturing of scale-up prototypes based on multiple layers prediction. *Rapid Prototyping Journal*, *31*(5), 1039–1052.
128. Thakur, A., Anadebe, V. C., Dagdag, O., Sharma, D., & Om, H. (2025). Fundamentals and recent applications of 3D, 4D, and 5D printing of polymer and its nanocomposites: Challenges and its future prospects. In *Polymer Nanocomposites for 3D, 4D and 5D Printing: Fundamental to Applications*, edited by Srikanta Moharana, Bibhuti B. Sahu, Santosh Kumar Satpathy, and Subhendu Chakroborty, 345–392. Singapore: Springer Nature. 10.1007/978-981-96-4214-4
129. Huang, W. J., & Li, D. R. (2025). In-situ monitoring and correction of geometric deviations in material extrusion additive manufacturing. *Rapid Prototyping Journal*, *31*(4), 612–628.
130. Alqutaibi, A. Y., Aljohani, R., Almuzaini, S., & Alghauli, M. A. (2025). Physical–mechanical properties and accuracy of additively manufactured resin denture bases: Impact of printing orientation. *Journal of Prosthodontic Research*, *69*(4), 461–471, https://doi.org/10.2186/jpr.JPR_D_24_00263.

131. El-Qemary, A., Kabbouri, I., Boutahari, S., & Chahbouni, M. (2025). Anticipation and correction of additive manufacturing geometric defects at the design stage. *Journal of Machine Engineering*, *25*, 74–88.
132. Rojek, I., Mikołajewski, D., Kempiński, M., Galas, K., & Piszcz, A. (2025). Emerging applications of machine learning in 3D printing. *Applied Sciences*, *15*(4), 1781.
133. Sartori, N., Sanchez, S. A., Oliveira, D., Hosney, S., Zoidis, P., Martin, W., ... Rocha, M. G. (2025). Flexural properties and fatigue limit of 3D-printed and milled resin-based materials. *Journal of Prosthodontics*, *34*(6), 626–634.
134. Wu, J., Shui, Y., Xie, C., Wu, Q., Yu, M., Luo, T., ... Yu, H. (2025). Effect of fabrication methods and number of supporting teeth on the surface accuracy and dimensional stability of implant surgical guides. *The Journal of Prosthetic Dentistry*, *133*(2), 530–e1.
135. Harun-Ur-Rashid, M., Foyez, T., Krishna, S. B. N., Poda, S., & Imran, A. B. (2025). Recent advances of silver nanoparticle-based polymer nanocomposites for biomedical applications. *RSC Advances*, *15*(11), 8480–8505.
136. Sitharaj, A., Arulmurugan, B., Balaji, D., & Idhayaraja, S. (2025). Additive manufacturing of titanium-based alloys by wire arc directed energy deposition (WA-DED): A review. In *Advances in Solid-State Welding and Processing of Metallic Materials*, edited by S. R. Vivekanandh and V. Veeraraghavan, 400–417. Boca Raton, FL: CRC Press/Taylor & Francis Group (Woodhead Publishing imprint). ISBN 9781040317938.
137. Malabadi, R. B., Chalannavar, R. K., Divakar, M. S., Swathi, K. K., Kamble, A. A., Karamchand, K. S., & Munhoz, A. N. R. (2025). Industrial *Cannabis sativa* (fiber or hemp): 3D printinghempcrete-a sustainable building material. *World Journal of Advanced Engineering Technology and Sciences*, *14*(02), 253–282.
138. Bourgi, R., Etienne, O., Holiel, A. A., Cuevas-Suárez, C. E., Hardan, L., Roman, T., ... Kharouf, N. (2025). Effectiveness of surface treatments on the bond strength to 3D-printed resins: A systematic review and meta-analysis. *Prosthesis*, *7*(3), 56.
139. Cong, B., & Zhang, H. (2025). Innovative 3D printing technologies and advanced materials revolutionizing orthopedic surgery: Current applications and future directions. *Frontiers in Bioengineering and Biotechnology*, *13*, 1542179.
140. Raj, R., Song, Q. C., & Juang, J. Y. (2025). Multi-material additive manufacturing of soft robotic systems: A comprehensive review. *Advanced Robotics Research* 2 (Early View): Article 202500064. https://doi.org/10.1002/adrr.202500064
141. Zhu, Y., Guo, S., Ravichandran, D., Ramanathan, A., Sobczak, M. T., Sacco, A. F., ... Song, K. (2025). 3D-printed polymeric biomaterials for health applications. *Advanced Healthcare Materials*, *14*(1), 2402571.
142. Ullah, A., Shah, M., Ali, Z., Asami, K., Ur Rehman, A., & Emmelmann, C. (2025). Additive manufacturing of ceramics via the laser powder bed fusion process. *International Journal of Applied Ceramic Technology*, *22*(3), e15087.
143. Huang, W., Garmestani, H., & Liang, S. Y. (2025). Microstructure evolution and the influence on residual stress in metal additive manufacturing with analytics. *Crystals*, *15*(5), 435.

144. Elsayed, M. S., El-Kouedi, A. Y., & Shokry, T. E. (2025). Effect of aging on the marginal fit of milled and printed zirconia crowns: An in-vitro study. *BMC Oral Health*, *25*(1), 221.
145. Masseni, F., Zumbo, A., Stumpo, L., Polizzi, G., Tetti, G., Lusetti, T., ... Maggi, F. (2025). Advancements in photocurable solid propellants for additive manufacturing. In *AIAA SCITECH 2025 Forum*, Orlando, Florida, USA, 6–10 January 2025, Paper AIAA 2025-2330, pp. 1–22. Reston, VA: American Institute of Aeronautics and Astronautics (AIAA). https://doi.org/10.2514/6.2025-2330
146. Bălțatu, M. S., Vizureanu, P., Sandu, A. V., Achitei, D. C., Nergis, D. D. B., Perju, M. C., & Nabiałek, M. (2025). Biocomposites: Materials, properties, and applications. In *Composite Materials: Science and Engineering*, edited by Petrică Vizureanu. London, UK: IntechOpen. https://doi.org/10.5772/intechopen.1010197
147. Kennedy, S. M., M, S., A, V., & K, A. (2025). 4D bioprinting for personalized medicine, innovations in implant fabrication and regenerative therapies. *Polymer-Plastics Technology and Materials*, *64*(12), 1–26.
148. Amponsah, O., Nopuo, P. S. A., Manga, F. A., Catli, N. B., & Labus, K. (2025). Future-oriented biomaterials based on natural polymer resources: Characteristics, application innovations, and development trends. *International Journal of Molecular Sciences*, *26*(12), 5518.
149. Rijwani, T., & Ramkumar, P. (2025). Thermal debinding for binder burnout in metal and ceramic processing. *Heat Transfer Engineering*, *46*(7), 615–626.
150. Li, Z., & Chang, L. (2025). Development of wear-resistant polymeric materials using fused deposition modelling (FDM) technologies: A review. *Lubricants*, *13*(3), 98.
151. Margvelashvili-Malament, M., Thompson, V., & Malament, K. A. (2025). Minimally invasive fixed prosthodontics: A narrative review. *Journal of Esthetic and Restorative Dentistry*, *37*(5), 1248–1254.
152. Shin, H., Kang, Y. J., Kim, H., & Kim, J. H. (2025). Effect of cement space settings on the marginal and internal fit of 3D printed definitive resin crowns. *The Journal of Prosthetic Dentistry*, *133*(3), 821–826.
153. Mirasadi, K., Yousefi, M. A., Jin, L., Rahmatabadi, D., Baniassadi, M., Liao, W. H., ... Baghani, M. (2025). 4D printing of magnetically responsive shape memory polymers: Toward sustainable solutions in soft robotics, wearables, and biomedical devices. *Advanced Science*, *13*(15), e13091.
154. Mehta, A., Vasudev, H., Sharma, A., & Fidan, I. (2025). Introduction to 3D printing–based healthcare applications: Robots and biomedical devices. In *Advances in 3D and 4D Printing of Medical Robots and Devices*, edited by Ankit Sharma and Ismail Fidan, 1–24. Amsterdam, Netherlands: Academic Press. https://doi.org/10.1016/B978-0-443-24861-0.00001-0
155. Aufa, A. N., Hassan, M. Z., & Ilyas, R. A. (2025). *Handbook of 3D Printing in Biomedical Applications*. 1st ed. Boca Raton, FL: CRC Press/Taylor & Francis Group, 2025. ISBN: 9781032860695.
156. Azab, A., Abdelhady, W. A., Elwakeel, E., Ashraf, M., Wally, R., Soliman, A., ... Abozaid, D. (2025). Systematic review and meta analysis of mechanical properties of 3D printed denture bases compared to milled and conventional materials. *Scientific Reports*, *15*(1), 29207.
157. Shehbaz, W., & Peng, Q. (2025). Selection and optimization of additive manufacturing process parameters using machine learning: A review. *Computer-Aided Design and Applications*, *22*(5), 893–911.

158. Alam, M. I., Kashyap, S., Balaji, P. G., Yadav, A. K., & Flora, S. J. S. (2025). 3D-printed medical implants: Recent trends and challenges. *Biomedical Materials & Devices*, *3*(2), 750–770.
159. Aktug Karademir, S., Atasoy, S., Akarsu, S., & Karaaslan, E. (2025). Effects of post-curing conditions on degree of conversion, microhardness, and stainability of 3D printed permanent resins. *BMC Oral Health*, *25*(1), 304.
160. Zhu, J., Wei, Y., Yang, G., Zhang, J., Liu, J., Zhang, X., ... Wang, J. (2025). Peptide-based rigid nanorod-reinforced gelatin methacryloyl hydrogels for osteochondral regeneration and additive manufacturing. *Nature Communications*, *16*(1), 7090.
161. Mahmud, M. S., Mkawe, P. N., & Nyagilo, V. O. (2025). A comprehensive analysis of non-planar toolpath optimization in multi-axis 3D printing: Evaluating the efficiency of curved layer slicing strategies. *Review of Applied Science and Technology*, *4*(02), 274–308.
162. Menagadevi, M., Nirmala, M., Thiyagarajan, D., & Somasundram, D. (2026). Biomaterials and their applications. *Biomedical Materials & Devices*, *4*(1), 46–61.
163. Santos, S., Matos, C., Duarte, I., Olhero, S. M., & Miranda, G. (2025). Effect of TPMS reinforcement on the mechanical properties of aluminium–alumina interpenetrating phase composites. *Progress in Additive Manufacturing*, *10*(2), 1187–1199.
164. Li, J., Yang, T., Liu, S., Shi, Y., & Zhao, K. (2025). A review on recent progress in light metal solid-phase additive manufacturing. *Materials Science and Technology*, *42*(7), 695–718. https://doi.org/10.1177/02670836251314968
165. Tu, J., Engelhardt, J. D., Barkow, M. H., Cruz Robles, J. C., Stanzione III, J. F., Chimenti, R. V., & Palmese, G. R. (2025). Additive manufacturing of polymeric gradient index optics via grayscale digital light processing vat photopolymerization technology. *ACS Applied Optical Materials*, *3*(2), 1405–1416.
166. Bouguermouh, K., Habibi, M., Laperrière, L., Li, Z., & Abdin, Y. (2025). Designing advanced 4D printing thermo-sensitive shape memory polymer blends for enhanced mechanical and shape memory performances. *Progress in Additive Manufacturing*, *10*(5), 3405–3424.
167. Sonkaya, E., Soygun, K., & Kahraman, Ö. C. (2025). Does cigarette smoke affect permanent resin-based crown materials? *Journal of Esthetic and Restorative Dentistry*, *37*(2), 586–594.
168. Pragana, J. P. M., Sampaio, R. F. V., Bragança, I. M. F., Silva, C. M. A., Nielsen, C. V., & Martins, P. A. F. (2025). Macro-scale finite element simulation of wire-arc additive manufacturing. *Proceedings of the Institution of Mechanical Engineers, Part L: Journal of Materials: Design and Applications*, *239*(5), 1013–1022.
169. Addess, A., Shilo, D., Glass, B., Paskovitch, Y., Burkhardt, C., Vogel, L., ... Faran, E. (2025). Additive manufacturing of Ni–Ti SMA by the sinter-based MoldJet method. *Shape Memory and Superelasticity*, *11*(1), 89–99.
170. Salehabadi, M., & Mirzadeh, H. (2025). 3D printing of polyester scaffolds for bone tissue engineering: Advancements and challenges. *Advanced Materials Technologies*, *10*(8), 2401522.
171. Mishra, V., Bharat, N., Kumar, V., Vellaisamy, M., & Veeman, D. (2025). Multi-objective optimization of 3D printed PLA/carbon fibre composite using a combined approach of gray relational analysis and ANOVA. *Journal of Thermoplastic Composite Materials*, 08927057251339246. https://doi.org/10.1177/08927057251339246

Chapter 12

Additive manufacturing and waste reduction

A revolution in the aerospace industry

Devesh Pant, Arun Kumar Chaudhary, Deepak Gupta, Deepanshu Arora, and Swasti Chaudhary

12.1 INTRODUCTION

The aerospace industry is undergoing a major change due to additive manufacturing (AM), or 3D printing. AM creates three dimensional (3D) objects by building them layer- by- layer from digital models, unlike traditional manufacturing, which removes material to shape an object. This technology is transforming aircraft design and production. Initially, AM was used for creating prototypes quickly ,but now it is used to create tools and components and in the production of complex, lightweight structures that increase aircraft performance and fuel efficiency. In traditional aerospace manufacturing, large amounts of raw materials are often wasted and it also requires large amounts of energy. There is a growing need to reduce waste due to strict regulations and potential cost savings, with strategies including reducing material use, recycling, and resource recovery. AM improves fuel efficiency by optimizing designs and reducing material waste, which is critical for both environmental compliance and cost savings. Engineers are using AM, not only to create tools and complex components but also to simplify assembly by merging multiple parts into a single structure, a feature that Boeing and Airbus are already taking advantage of in their manufacturing processes (Wikipedia Contributors 2025). In fact, Boeing uses AM to produce titanium parts for its 787 Dreamliner (Dehoff et al. 2013), cutting material waste by nearly 50%. While NASA and SpaceX are exploring AM for rocket engine components (Chiu 2020). SpaceX's successful tests of 3D-printed SuperDraco engines are already paving the way for more affordable and efficient rocket designs.

12.2 AM IN AIRCRAFT MANUFACTURING

Additive Manufacturing (AM) is transforming aircraft design, production and performance (Li et al. 2024). While the aerospace sector has used AM technologies for decades, its initial applications were limited in creating prototypes quickly. Today, AM is used for the production of tools, components, and functional parts. Thus, offering significant advantages in terms

 DOI: 10.1201/9781003607052-12

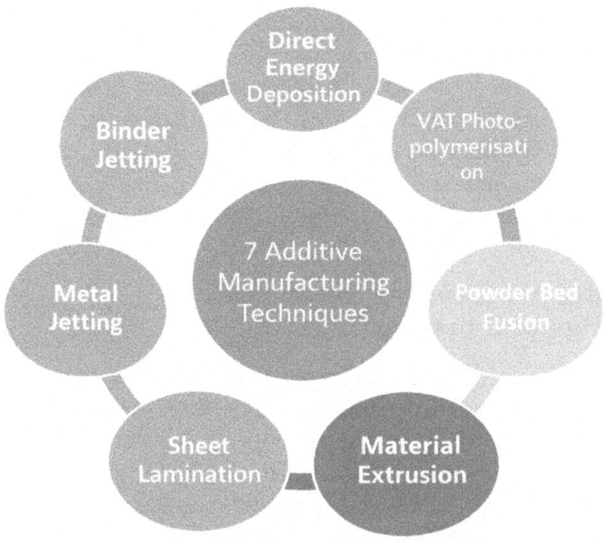

Figure 12.1 The various processes of additive manufacturing. (Source: Alami et al. 2023.)

of cost, efficiency, and design flexibility. The Figure 12.1 shows some of the AM techniques which are used in the industry.

12.2.1 3D printing in commercial aircraft

AM is transforming commercial aircraft manufacturing by producing both critical and non-critical components. These include nozzles, engine parts, and even brackets and fixtures, which allow for more efficient and cost-effective production. Major aerospace companies like Boeing and Airbus use AM to streamline production and cut costs. Boeing has produced over 20,000 parts with AM, saving $2–3 million per airplane by switching to titanium-alloy parts demonstrating that AM can lead to long-term cost savings (Li et al. 2024). While Airbus uses AM to manufacture lightweight metallic brackets and bleed pipes and leveraged AM to incorporate lightweight, high-performance components into its aircraft, reinforcing the technology to revolutionize aerospace design.

12.2.2 Lightweight components and structural optimization

In the aerospace industry, materials must be lightweight yet strong to meet safety standards, reduce emissions, and improve fuel efficiency. AM produces the components that meet these requirements. The technology creates strong complex geometries with minimum material, which is particularly valuable in aerospace where weight reduction is critical. The AM builds parts with layer-by-layer approach which fabricates highly accurate and optimized parts. As a result, the components are both lighter and stronger, with the weight reduced by of 40%–80%. Additionally, AM facilitates the redesign of components for specific functions. It eliminates unnecessary material through advanced geometries (Gupta 2024). For example, a redesigned aerospace bracket weighing just 326 g was developed using AM, which shows the technology's potential for weight reduction and performance enhancement.

12.2.3 Rapid prototyping for aircraft design and testing

AM was initially used in the aerospace industry for creating prototypes quickly, which were non-functional parts used to verify simulation models and test the design of aircraft components. Prototypes were very useful in testing aerodynamics, such as airflow around airfoils and fuselage components by finding design flaws and functional issues which did not appear in digital models. Prototypes may or may not be made from the final part material. However, they still need to be rigid and accurate enough to provide reliable test results. AM processes such as stereolithography (SLA) and fused deposition modeling (FDM) are best for making these prototypes. They are cost-effective polymer-based materials, they reduce lead times, and fast design change. This makes AM a valuable tool for the development cycle. For example, PolyJet technology was used to fabricate prototypes for testing various wing designs in unmanned aerial vehicles (UAVs), which tested the propulsion, aerodynamics, and structural performance. The FDM process is commonly used for producing simple prototypes in the early stages of design. Which provides a fast and affordable way to check concepts before moving to high quality production. The Table 12.1 compares the different materials used in AM for aerospace, which highlight

Table 12.1 Comparison of materials used in additive manufacturing for aerospace industries

Material	*Aerospace application*	*Advantages*	*Disadvantages*
Polymers	Prototyping and concept development	Low cost and ease of processing	Limited mechanical strength and load-bearing capacity
	Interior components and cabin parts	Lightweight, suitable for interior components	Limited thermal stability and chemical resistance
	Non-structural parts and ducting systems	Excellent design flexibility and complexity	Limited dimensional accuracy and potential for warping or distortion during printing

(Continued)

Table 12.1 (*Continued*) Comparison of materials used in additive manufacturing for aerospace industries

Material	*Aerospace application*	*Advantages*	*Disadvantages*
	Lightweight drone components and UAVs	Good impact resistance and vibration damping	Limited recyclability and environmental concerns
Metals and alloys	Engine components and turbines	High strength-to-weight ratio	High material and processing costs
	Structural components and airframes	Excellent mechanical properties	Limited design complexity and intricate features
	High-temperature applications and aerospace structures	High thermal conductivity	Limited availability of AM-grade materials for high-performance applications
	Propulsion systems and rocket components	Good wear resistance and fatigue resistance	Potential for microstructural defects in printed parts
	Satellite components and communication systems	High resistance to extreme temperatures and harsh environments	Post-processing may be required to achieve desired mechanical properties
Ceramics	Turbine blades and combustion systems	High-temperature resistance and thermal stability	Limited design complexity
	Propulsion systems and rocket nozzles	Excellent mechanical properties	High processing temperatures required for sintering
	Ceramic matrix composites for aerospace structures	Low density, lightweight	Cost and availability of specialized ceramic powders
	Satellite components and communication systems	Good chemical inertness and resistance to corrosion	Challenging to achieve dense and void-free prints due to high processing temperatures

Source: Alami et al. (2023).

their strengths and limitations to help manufacturers choose the best fit for each application and offers information on the various materials used in AM processes. It shows their properties, advantages, and limitations, which makes it easier to choose the right materials for specific aerospace applications. AM can create prototypes, structural components or can do rapid functional testing. This helps manufacturers to change their designs, reduce production costs, and enhance the overall performance and reliability of aerospace components just by selecting the right material of required property (Alami et al. 2023).

12.3 MATERIAL EFFICIENCY AND WASTE REDUCTION IN AEROSPACE AM

AM helps protect the environment by designing parts that use less material and create less waste. Research from many sources (Froes, Boyer, and Dutta 2017) shows that AM can reduce primary energy use and CO_2 emissions by optimizing designs and cuts down waste by selectively depositing material, which reduce waste compared to traditional methods. AM uses a layered approach for the construction of components, which use only the necessary material, further reducing waste production and increase the efficiency (Al-Lami, Hilmer, and Sinapius 2018). Table 12.2, "Metal Additive Manufacturing (MAM) Processes and Their Characteristics" presents data, which estimates how additive manufacturing (AM) leads to significant material reductions in various aerospace components. This data shows that AM processes like electron beam melting (EBM), selective laser melting (SLM) and other Metal AM processes save material and make aircraft lighter. This improves the fuel efficiency and reduces the cost.

12.3.1 Sustainable material selection for aerospace AM

AM uses lightweight materials and smart designs to make aircraft parts that are lighter, which boosts fuel efficiency and cuts down carbon emissions. The design of aerospace components focuses on creating complex structures with minimum material. Materials like titanium alloys and nickel-based alloys, such as Ti6Al4V (Ti64), are commonly used because they can handle high heat and resist corrosion, ensure safety in demanding aerospace industry (Monteiro et al. 2022). Aluminum alloys have lower toughness. But, have gained interest due to their lightweight potential, especially in understanding potential defects and quality in aluminum parts. AM can reduce waste and energy consumption during the manufacturing process, conserve time and energy throughout various production stages. Table 12.3 shows how AM helps reduce the weight of critical aerospace components, improving fuel efficiency and cutting operational costs. The data shows how using materials like titanium, aluminum, and nickel alloys with AM leads to lighter and stronger designs. By decreasing the mass of critical aerospace components, AM improves fuel efficiency and reduces operational costs. This table highlights the environmental and economic advantages of AM by quantifying the reduction in raw material consumption across different aerospace systems.

$$\text{Reduction } \% = \left(1 - \left(\rho_{\text{alternative}} / \rho_{\text{traditional}}\right)\right) \times 100 \tag{12.1}$$

Table 12.2 Metal additive manufacturing (MAM) processes and their characteristics

Process	*Energy source (W)*	*Process efficiency (%)*	*Material type*	*Inert gas (IG) or vacuum*	*Weld pool size (mm²)*	*Deposition speed (kg/h)*	*Comments*
Selective laser sintering (SLS)	Laser	-	Powder	-	-	-	Similar to SLM, but less used.
Selective laser melting (SLM)	Laser 100–1,000	2%–5%	Powder	IG: argon	0.3	0.1–0.18	Suitable for small and complex parts. Smoother surface than EBM but with higher cooling rates. More mechanical defects and residual stress than EBM, requiring thermal treatment.
Electron beam melting (EBM)	Electron beam 3,500	15%–20%	Powder	Vacuum at constant high temp.	1	0.2–0.36	High accuracy with density and mechanical properties close to cast material. Lower yield stress but higher strain than SLM. Usually, no thermal treatment required. Lower fatigue resistance but better efficiency than SLM.
Laser metal deposition (LMD)	Laser 500–3,000	2%–5%	Wire	IG: argon	Smaller than WAAM	-	Lower power efficiency for highly reflective alloys (40%–95% reflectance). Better dimensional properties than WAAM but slower deposition rate and efficiency. Can be used to repair high-value parts.
Electron beam free form fabrication (EBFFF)	Electron beam	-	Wire	Vacuum	2.3–7 < 18	-	High energy efficiency (~90%) even with reflective alloys.

(Continued)

Table 12.2 *(Continued)* Metal additive manufacturing (MAM) processes and their characteristics

Process	*Energy source (W)*	*Process efficiency (%)*	*Material type*	*Inert gas (IG) or vacuum*	*Weld pool size (mm^2)*	*Deposition speed (kg/h)*	*Comments*
Wire arc additive manufacturing (WAAM)	Electric arc 2,000–4,000	70%	Wire	IG: argon or similar	0.5–4 (up to 10)	-	High material (~90%) and energy (~70%) efficiency. Cost-effective for large and less complex parts. Rough surface, lower accuracy, requiring more surface finishing. Cold-high pressure rolling may be used for large components.
Wire laser additive manufacturing (WLAM)	Wire	-	Wire	IG: argon or helium	-	-	Better dimensional properties than WAAM but lower deposition rate and material efficiency.

Source: Monteiro et al. (2022).

Table 12.3 Estimated mass reductions by material and component system through AM replacement

Component system	*Category*	*Representative component*	*Traditional material*	*Lightweight alternative*	*Documented benefit*	*Potential mass reduction (if quantified)*
Primary structure	Fuselage	Fuselage panels	2024-T3 Al alloy	Al-Li (2060-T8)	"7% and 14% weight reduction compared to 2524 and 2014 alloys, respectively"	7%–14% (vs. conventional Al)
	Wings	Upper wing skins	7075 Al alloy	7085 Al alloy	"Excellent mechanical properties (YS=504 MPa) and better damage tolerance	-
Engine	Cold section	Compressor blades	Ti-6Al-4V	Ti-6Al-4V/SiC fiber composite	"Higher specific strength and creep resistance"	Not quantified
	Hot Section	Turbine blades	Ni-based superalloys (IN792)	AM-processed Ni superalloys (SLM)	"AM enables fabrication of complex geometries". No mass reduction data provided.	-
Landing gear	Structural	Landing gear beams	300M Steel (YS: 1689 MPa)	Ti-10V-2Fe-3Al (β-Ti alloy)	"40% lower density than steel").	~30%–40% (density-based estimate)
Furnishings	Functional	Seat components	7XXX Al alloys	Mg alloys (WE43A)	"35% lower density than Al". Limited to non-critical parts due to corrosion issues.	~20%–30% (density-based estimate)
Component system	Category	Representative component	Traditional material	Lightweight alternative	Documented benefit	Potential mass reduction (if quantified)
Primary structure	Fuselage	Fuselage panels	2024-T3 Al alloy	Al-Li (2060-T8)	"7% – 14% weight reduction compared to 2524 and 2014 alloys, respectively".	7%–14% (vs. conventional Al)

Source: Monteiro et al. (2022) and Gloria et al. (2019).

Note: "equation 2.1" is a density-based mass reduction formula was used for some comparison of materials for lightweight potential in aerospace and other industries in the Table 12.3.

12.3.2 Recycling and reusability of am powders and filaments

Researchers are exploring ways to recycle wasted powders, and the recent advancements have emerged in recycling technologies are very promising. Some of these efforts include reusing scrap materials and even developing customized packaging designed for recycling in space (Wikipedia Contributors 2024) Figure 12.2 shows the hierarchy of waste management in the aerospace industry

12.3.3 Reducing material waste through design optimization

AM can easily optimize the design, which allows us to redesign the components for specific functions and eliminate the unnecessary material through complex geometries in Lattice structures. AM technologies like selective laser melting (SLM) and fused deposition modeling (FDM) can create these strong and complex geometries like honeycomb structures with high precision. In AM, multiple components are integrated into a unified framework, which enhances aircraft performance and increases fuel efficiency by reducing weight without compromising structural strength (Wikipedia Contributors 2025). In aerospace manufacturing, the buy-to-fly (BtF) ratio measures the amount of raw material required to produce a final part (Chiu 2020). Traditional manufacturing methods generated a high BtF ratio and a significant portion of the raw material is wasted during machining or shaping. The average BtF ratio for aircraft components typically ranges from 8

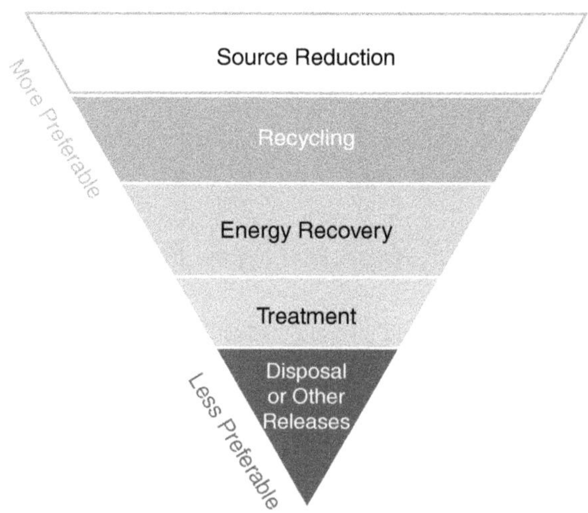

Figure 12.2 Waste management hierarchy. (Source: Gaona 2018.)

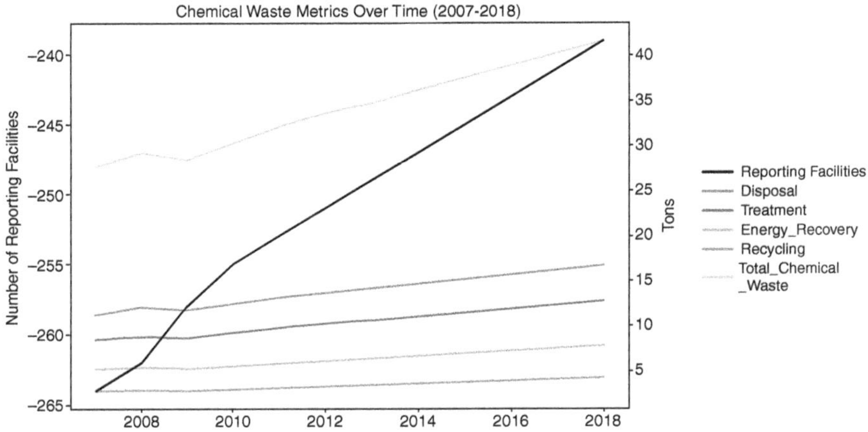

Figure 12.3 Toxic release inventory (TRI) chemical waste metrics for aerospace industry (2007–2018). (Source: U.S. Environmental Protection Agency 2025.)

to 12, with some complex parts reaching as high as 30. Such strategies for eliminating the use of hazardous materials have generated more hazardous waste. Aerospace facilities now use source reduction activities to reduce scrap metal generation, especially with materials such as chromium, nickel, and copper, which deals waste at its source, as it is a waste management practices. Figure 12.3 shows how waste generation trends have evolved in the aerospace sector. The data showed the adoption of additive manufacturing (AM) has played a role in reducing chemical waste through design optimization, reducing waste from source and by enhancing material efficiency (U.S. Environmental Protection Agency 2025). The comparison of waste metrics over a decade shows that the continuous efforts of aerospace sector by using AM in their processes has minimized environmental impact while upholding high-performance standards. Thus, they have tackled waste management challenges and promoting sustainable manufacturing practices.

12.4 AM IN AEROSPACE MAINTENANCE AND SPARE PARTS PRODUCTION

Additive manufacturing (AM) is transforming aircraft maintenance by offering efficient and sustainable solutions for repairing critical components. AM technologies are being used in many aerospace applications like direct part manufacturing, rapid tool and prototypes making, and parts restoration. For instance, laser-based AM processes allow for the precise repair of turbine blades, extending the life of critical aircraft components (Gupta 2024).

12.4.1 3D printing for aircraft spare parts

AM improves spare part supply chains by making them more adaptable and reduces lead times. It decentralizes production, bringing it closer to customers, allowing companies to respond quickly and enhance service levels, ultimately lowering aircraft building and operating costs (Ivkić et al. 2024). For example, Airbus delivered the first of GE's LEAP engines with integrated 3D-printed fuel nozzles. It achieved a 25% weight reduction by reducing the number of parts from 20 to 1 and shorter assembly times. Airbus is actively utilizing AM to reshape its supply chain, allowing for on-demand parts delivery. Companies like Materialise and EOS have pioneered the 3D printing of flight-ready selective laser sintering (SLS) parts for Airbus. Additionally, mathematical models can be developed to optimize the deployment of 3D printers in specific locations, reducing total operational costs even for large supply chains. Additionally, AM technologies, including DED (Direct Energy Deposit), allow for the repairing and restoring of both metallic and non-metallic aircraft parts, which allows us to reuse them instead of scrapping. This saves costs and reduces environmental impact. Geometry restoration, like DED, is a cost-effective restoration method of missing or damaged aircraft components. For example, a T700 blisk was commercially repaired from erosion damage using DED technology, while a low-wattage repair method effectively restored titanium components with minimal distortion. Compressor blade tips have also been repaired using AM technology, enhancing engine performance.

12.4.2 On-demand manufacturing of UAVs and drones

AM is also used to fabricate complete drones or specific components. With processes like the PolyJet process are used for the rapid testing of UAV wings. AM in UAV manufacturing reduces material usage and saves time while maintaining or improving structural strength through complex structures like honeycombs (Romaniw and Bras 2012). A fixed-wing UAV can be printed in about 3 hours where the conventional methods can take upto 4 hours. Studies on the design and layout of 3D-printed UAV wings focus on achieving optimal lightweight structures and load distribution.

12.4.3 Reducing downtime and maintenance costs with AM

Additive manufacturing reduces maintenance downtime, which is crucial for aircraft operations. AM processes like DED can restore complex geometries with consistent material properties and avoid residual stress, where stereolithography (SLA), which is a process that uses UV light to cure photopolymer resin can fabricate repair kits quickly with some polymer-based components in high-stress environments (de Brito et al. 2020). Compressor blade tips have also been repaired using AM technology, which enhanced engine performance. Structural restoration techniques improve the integrity of components by repairing cracks and corrosive damage using solid-state processing (SPD). Titanium SPD has been successfully used to repair hydraulic lines and overlapping joints to prevent further corrosion. AM processes like laser metal deposition (LMD) and directed metal deposition (DMD) are used for aircraft maintenance (Huang et al. 2016). They repair the critical components like compressors, turbines, and combustor coatings. These processes effectively restore high-performance parts that would otherwise be scrapped, enhancing operational efficiency. The aerospace suppliers like Eaton Aerospace are using AM capabilities by developing high-performance materials such as titanium, Inconel, and stainless steels, which are compatible with Airbus's safety standards. The print-on-demand approach not only saves on storage space but also reduces production costs. Furthermore, using mobile additive manufacturing as an on-site technology for space exploration allows for the local production of required components by utilizing new extraterrestrial materials, eliminating the need to transport all materials from Earth. Figure 12.4 shows a range of 3D-printed part produced using additive manufacturing (AM) technologies. It shows how AM is a quick and cost-efficient fabrication method for critical parts and their replacement parts (Getachew, Shiferaw, and Ayele 2023). This ability to optimize maintenance, especially for low-volume applications, can reduce lead times.

Figure 12.4 3D printed liquid hydrogen and liquid oxygen turbopumps at NASA MSFC3D. (Source: Blakey-Milner et al. 2021.)

12.5 CHALLENGES AND LIMITATIONS OF AM IN AEROSPACE

Despite the progress and benefits of AM, the aerospace sector faces various obstacles in achieving extensive use of this technology. Systematic attention and innovative solutions are necessary to address crucial concerns. The aerospace industry is burdened with regulations and standards established due to past uncertainties in production and design (Getachew, Shiferaw, and Ayele 2023). Many developmental methods can become too costly and time-consuming, which discourages companies from pursuing further development. One major challenge is the need for high-quality materials that meet the strict safety and performance standards required by the aerospace industry. AM technologies must ensure consistent dimensional accuracy and process repeatability to comply with aerospace safety standards, where even minor deviations can compromise structural integrity. The complexity of AM processes needs skilled personnel and specialized training, which further adds to operational costs. There is a significant need for holistic development, requiring the aerospace industry to collaborate with agencies to create the necessary standards for adopting and implementing AM.

12.5.1 Technical challenges in large-scale aerospace AM

The size limitation of additive manufacturing, like dimensions of the printing equipment, is one of the biggest hurdles in the aerospace industry (Blakey-Milner et al. 2021). As it can restrict the size of components that can be manufactured like wings, fuselage sections and engine parts, which often exceed the build capacity of traditional 3D printers. This needs an innovative approach to manufacture large-scale structures, at the same time maintain consistent quality, which is a considerable challenge as the fluctuations in material characteristics can lead to dimensional inaccuracies, porosity, and defects. These issues are critical due to the precision and reliability required in aerospace applications. AM can create complex shapes and lightweight constructions, but suitable materials that meet the aerospace standards for strength, durability, heat resistance, and fatigue performance are not easily available.

The aerospace components produced through AM require post-processing with precise finishing to meet stringent quality and performance criteria. However, the manual post-processing is time-consuming and prone to human error. Issues like surface roughness, which can hinder the series production is a common issue, especially for new companies. There are some other challenges like geometric repeatability, residual stress, and high surface roughness, which are more common in applications which requires high dimensional precision and fatigue resistance. The anisotropic properties of materials produced through AM can also complicate performance expectations. As detecting defects in complex internal structures is difficult with limited correction methods.

12.5.2 Certification and regulatory barriers in aviation and space industries

The aerospace industry faces a significant certification challenge due to a lack of current standards and certification processes for AM. The current certifications, design guidelines and regulations were created for the traditional manufacturing methods and hence, the rapid advancements in AM have outpaced them, which cause the industry leaders to hesitate in further research of AM. The customization of machines makes the certification process more complex. This led to concern that the development of future machines, alloys or processes may not be possible until the industry matures in terms of materials, machines, and technologies in aerospace industry. This has caused the original equipment manufacturers (OEMs) to be reluctant to fully embrace AM (Poyraz and Kuşhan 2019). The strict regulations and standards in the aerospace sector, established due to past uncertainties, can make developmental methods costly and lengthy, discouraging innovation. As the aerospace industry enters a digital age, there is an opportunity to revisit these stringent guidelines that may inhibit the adoption of new processes.

12.5.3 Economic and production constraints

The AM offers various benefits. However, the cost-effectiveness remains a significant challenge in aerospace engineering. The initial investment in high-end AM equipment and materials and huge operational costs due to the need for skilled personal with specialized training to perform the complex AM processes, creates a barrier for some companies. The stringent quality control standards for reliability and safety, the certification process for aerospace-grade parts produced through AM is time-consuming and expensive (Blakey-Milner et al. 2021). Limitations in current AM techniques, such as material properties and production speed, may hinder widespread adoption for mass production in the aerospace industry. The output of AM processes is currently low, which makes these technologies less suitable for high-volume production. The development of AM in aerospace is challenging, and systemic challenges, like the need for necessary tools, partnerships and funding, make it even more costly. Companies often lack best practices, expertise, machines and software to facilitate adoption. The aerospace OEMs are hesitant to allocate personnel to understand and develop AM capabilities. The Figure 12.5 visually represents the challenges.

Figure 12.5 Disadvantages of additive manufacturing in the aerospace industries.

12.6 FUTURE OF AM IN THE AEROSPACE INDUSTRY

Additive manufacturing (AM) technologies have already made a significant impact on the aerospace industry, and the future of this technology in this field is promising. The ability to create complex geometries and lightweight structures with high accuracy and precision has led to the production of innovative aircraft components and structures. As AM continues to mature into a robust production process, it is predicted to become a breakthrough technology in aerospace. If the potential of AM components is fully realized, airplane fuel consumption could be reduced by as much as 6.4% (Huang et al. 2016).

12.6.1 Advancements in AM technologies for aerospace applications

Continuous research and development in this technology is focused on optimizing material qualities, enhancing process control, and increasing the scalability of AM methods (Froes, Boyer, and Dutta 2017). Advancements in materials, processes, and machines are being made constantly, expanding the design space to address fit, form, and function requirements. This ongoing pursuit of innovation holds significant potential for the future of additive manufacturing. By 2050, the cumulative energy and carbon savings achievable from using AM parts in airplanes could be substantially greater with earlier adoption.

12.6.2 Integration of AI and simulation in AM design

Future efforts will focus on further investigating robotic AM processes and system integration to explore the full capabilities of mobile and additive manufacturing for smart manufacturing systems (Li et al. 2024). This includes developing a multi-level intelligent decision-making model for robotic AM systems that enables on-demand, on-site, and cloud-computing-controlled manufacturing of customized products. Vast improvements in flexible design and development methods will be based on simulation and generative algorithms, with weight reduction and part consolidation being significant factors through the use of topology optimization technologies and advanced simulation and design packages.

12.6.3 The role of AM in sustainable space exploration

Additive manufacturing is expected to have a significant impact on the production of space vehicles and satellites. The lightweight and durable structures produced using AM will allow for the creation of more efficient and cost-effective space vehicles (Watts et al. 2024). The ability to produce complex geometries will enable the development of more advanced and sophisticated satellites. One strategy for space exploration involves using AM as an on-site manufacturing technology, allowing for the local production of required components by utilizing new extraterrestrial materials instead of transporting all materials from Earth

12.7 CONCLUSION

Additive manufacturing has demonstrated significant advancements within the aerospace sector, facilitating unparalleled design capabilities, streamlined production procedures, and environmentally conscious manufacturing methodologies. AM enables the quick creation of prototypes and small batches of unique parts, providing greater flexibility in product development and improvement. Hence, AM becomes an appealing solution for optimization of the operations and promoting sustainability in the aerospace industry.

The process of harnessing the revolutionary capabilities of additive manufacturing is still ongoing and hence, it requires collaborative efforts from researchers, engineers and industry experts to overcome the remaining challenges. Ongoing research and development will address the obstacles associated with AM. By promoting innovation, fostering collaboration across various fields, and prioritizing quality and safety, the aerospace industry can effectively utilize additive manufacturing to achieve unprecedented levels of excellence. As barriers to mass adoption of AM begin to fade with technological advancements, more aerospace companies are expected to adopt additive manufacturing in the future.

REFERENCES

Alami, A. H., A. G. Olabi, A. Alashkar, S. Alasad, H. Aljaghoub, H. Rezk, and M. A. Abdelkareem. 2023. "Additive Manufacturing in the Aerospace and Automotive Industries: Recent Trends and Role in Achieving Sustainable Development Goals." *Ain Shams Engineering Journal* 14: 102516. https://doi.org/10.1016/j.asej.2023.102516.

Al-Lami, A., P. Hilmer, and M. Sinapius. 2018. "Eco-Efficiency Assessment of Manufacturing Carbon Fiber Reinforced Polymers (CFRP) in Aerospace Industry." *Aerospace Science and Technology* 79: 669–678. https://doi.org/10.1016/j.ast.2018.06.020.

Blakey-Milner, B., et al. 2021. "Metal Additive Manufacturing in Aerospace: A Review." *Materials & Design* 209: 110008. https://doi.org/10.1016/j.matdes.2021.110008.

Chiu, B. W. 2020. *Additive Manufacturing Applications and Implementation in Aerospace*. Master's thesis, Massachusetts Institute of Technology. https://hdl.handle.net/1721.1/126950.

deBrito, F. M., G. da Cruz Júnior, E. M. Frazzon, J. P. Basto, and S. G. S. Alcalá. 2020. "Design Approach for Additive Manufacturing in Spare Part Supply Chains." arXiv. https://doi.org/10.48550/arXiv.2011.10572.

Dehoff, R. R., C. Duty, W. Peter, Wei Chen, C. Blue, Cory Tallman. 2013. "Case Study: Additive Manufacturing of Aerospace Brackets." *Advanced Materials & Processes* 171. https://api.semanticscholar.org/CorpusID:138118439.

Froes, F. H., R. Boyer, and B. Dutta. 2017. "Additive Manufacturing for Aerospace Applications, Part I." *AM&P Technical Articles* 175, no. 5: 36–40. https://doi.org/10.31399/asm.amp.2017-05.p036.

Gaona, S. D. 2018. "The Utility of the Toxic Release Inventory (TRI) in Tracking Implementation and Environmental Impact of Industrial Green Chemistry Practices in the United States." In *Green Chemistry*, edited by Hosam El-Din M. Saleh and Martin Koller. London, UK: IntechOpen. https://doi.org/10.5772/intechopen.70716.

Getachew, M. T., M. Z. Shiferaw, and B. S. Ayele. 2023. "The Current State of the Art and Advancements, Challenges, and Future of Additive Manufacturing in Aerospace Applications." *Advances in Materials Science and Engineering* 2023: 1–13. https://doi.org/10.1155/2023/8817006.

Gloria, A., R. Montanari, C. Rinaldi, and T. Russo. 2019. "Alloys for Aeronautic Applications: State of the Art and Perspectives." *Metals* 9, no. 6: 662. https://doi.org/10.3390/met9060662.

Gupta, A. 2024. "Advantages and Challenges of Additive Manufacturing: A Breakthrough in Aerospace Manufacturing." *International Journal of Research in Aeronautical and Mechanical Engineering* 12, no. 3: 1–14. https://doi.org/10.5281/zenodo.10847767.

Huang, R., M. Riddle, D. Graziano, J. Warren, S. Das, S. Nimbalkar, J. Cresko, and E. Masanet. 2016. "Energy and Emissions Saving Potential of Additive Manufacturing: The Case of Lightweight Aircraft Components." *Journal of Cleaner Production* 135: 1559–1570. https://doi.org/10.1016/j.jclepro.2015.04.109.

Ivkić, I., T. Buhmann, B. List, and C. Gnauer. 2024. "Towards a Cost-Benefit Analysis of Additive Manufacturing as a Service." arXiv preprint arXiv:2403.18882. https://doi.org/10.48550/arXiv.2403.18882.

Li, Y., J. Park, G. Manogharan, F. Ju, and I. Kovalenko. 2024. "A Mobile Additive Manufacturing Robot Framework for Smart Manufacturing Systems." *ASME 2024 International Manufacturing Science and Engineering Conference, MSEC2024*. https://doi.org/10.1115/MSEC2024-125413.

Monteiro, H., G. Carmona-Aparicio, I. Lei, and M. Despeisse. 2022. "Energy and Material Efficiency Strategies Enabled by Metal Additive Manufacturing—A Review for the Aeronautic and Aerospace Sectors." *Energy Reports* 8: 298–305. https://doi.org/10.1016/j.egyr.2022.01.035.

Poyraz, O., and M. C. Kuşhan. 2019. "Design for Additive Manufacturing with Case Studies on Aircrafts and Propulsion Systems." *Scientific Research and Education in the Air Force-AFASES* 21: 23. https://doi.org/10.19062/2247-3173.2019.21.23.

Romaniw, Y., and B. Bras. 2012. "Survey of Common Practices in Sustainable Aerospace Manufacturing for the Purpose of Driving Future Research." In *Leveraging Technology for a Sustainable World*, edited by D. Dornfeld and B. Linke, 82–95. Berlin, Heidelberg: Springer. https://doi.org/10.1007/978-3-642-29069-5_82.

U.S. Environmental Protection Agency. 2025. "Aerospace Manufacturing Sector Pollution Prevention (P2)". Accessed February 22, 2025. https://www.epa.gov/toxics-release-inventory-tri-program/aerospace-manufacturing-sector-pollution-prevention-p2.

Watts, H., R. Premo, W. Huberty, C. Bounds, and H.-G. Kim. 2024. "Experimental Investigation of the Structural Performance of Composite Structures Produced Using Additive Manufacturing." *AIAA Scitech 2024 Forum 2288*. https://doi.org/10.2514/6.2024-2288.

WikipediaContributors. 2024. "Space Manufacturing." Wikipedia, The Free Encyclopedia. Last modified December 30, 2024. Accessed February 22, 2025. https://en.wikipedia.org/wiki/Space_manufacturing.

WikipediaContributors. 2025. "3D Printing." Wikipedia, The Free Encyclopedia. Last modified February 16, 2025. Accessed February 22, 2025. https://en.wikipedia.org/wiki/3D_printing.

Chapter 13

The role of IoT in modern metal additive manufacturing

Evolution, current innovations, and future prospects

Shahroz Akhtar Khan

13.1 INTRODUCTION

Additive manufacturing (AM) offers special benefits that enable the production of complex personalized products through its capability of on-demand production while minimizing the waste [1]. Various process parameters in AM processes enable users to enhance product properties [2–4]. Today, MAM stands as a crucial modern manufacturing technology which allows users to construct detailed metal machine parts while minimizing the manufacturing time [5]. Production techniques like selective laser melting (SLM), in addition to laser metal deposition (LMD), together with cold spray additive manufacturing (CSAM), enable industries, especially aerospace and biomedical engineering, to produce small parts rapidly through prototypes while reducing manufacturing cost [6,7]. Simultaneously, the integration of IoT technologies has witnessed a transformation in the manufacturing domain globally. IoT adds to real-time machining-data acquisition, remote monitoring of fabrication processes, and closed-loop control, which have been essential in improved repeatability, reliability, and efficiency of MAM processes [8,9].

Recent research has demonstrated the benefits of adopting IoT in MAM; for example, real-time machining-data acquisition permits in-situ monitoring of machining process variables [10,11]. Furthermore, recent developments in ML have opened up new possibilities for optimizing process parameters, fault prediction, with the end objective of improving the quality of fabricated metal machine parts [12–14]. Despite these advancements, paramount concerns that are not yet resolved include a lack of standardized protocols for IoT, risks in the realm of cybersecurity, and scalability challenges [15,16]. This paper seeks to render a comprehensive investigation of IoT-enabled applications in MAM through the evaluation of the past developments, current breakthroughs, and future prospects.

 DOI: 10.1201/9781003607052-13

13.2 HISTORICAL DEVELOPMENT OF IOT IN MAM

Reference can be made to Figure 13.1 for various stages in the historical development of integration of IoT in MAM processes.

13.2.1 Early sensor integration

The integration of IoT concepts into MAM began in the early 2010s when studies started to test out sensor-based monitoring of important process parameters [17,18]. Researchers initially relied on thermocouples alongside pyrometers and infrared cameras to track temperature changes [19]. The researchers studied layer adhesion dynamics together with melt pool behavior through offline data analyses [20]. Static temperature monitoring systems were developed as a research discovery to predict SLM defects, which opened doors toward future IoT applications [21].

13.2.2 Emergence of real-time monitoring and data analytics

Technological advances enabled sensors to send live information over transmission channels. Early in the 2010s, MAM systems began

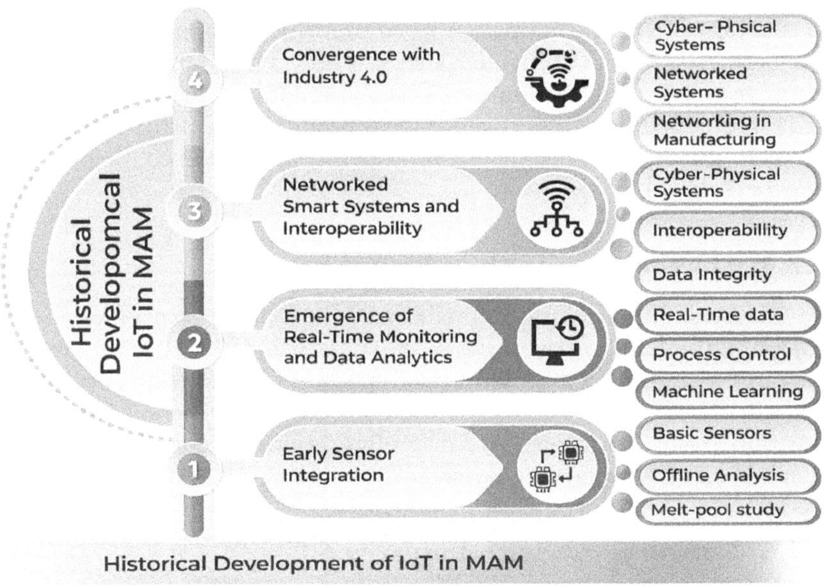

Figure 13.1 Development stages of IoT applications in MAM.

integrating low-latency data protocols and wireless sensor networks (WSNs) according to research [22,23]. New developments enabled real-time monitoring of processes, which let users pinpoint and handle process irregularities [24]. Process control was further enhanced by fusing real-time sensor data analysis with ML methods [25,26]. The analysis confirmed that IoT structures functioned as potential solutions to enhance manufacturing quality by implementing predictive maintenance and adaptive control paradigms.

13.2.3 Networked smart systems and interoperability

The MAM system received its networked smart capabilities through integration with different sensors. Through modern wireless communication technology, the system presents an improvement that enables the implementation of multiple sensors and devices for collecting and analyzing real-time data system-wide [8,9]. The strong networking protocols assist WSNs and edge computing systems to collaborate in synchronized responses to operational changes, which improves predictive maintenance and quality assurance alongside adaptive control [22,23].

The achievement of interoperability among diverse devices was a notable milestone during this time period. The early initiatives in MAM highlighted the need for defined protocols to facilitate smooth communication. The introduction of standards such as open platform communications unified architecture (OPC UA) provided a robust and dependable foundation for data transmission, addressing concerns about data integrity and cybersecurity [15,16]. This move improved system performance while also enabling genuinely integrated cyber-physical systems, which are characteristics of Industry 4.0 [27].

13.2.4 Convergence with Industry 4.0

The Industry 4.0 revolution emphasized the significance of data-driven decision-making and networking in manufacturing [27]. In order to create integrated cyber-physical systems (CPS) to interact with enterprise systems, researchers started integrating IoT modules into MAM setups [28,29]. The transition from standalone sensor configurations to complete networked systems was a major turning point.

13.2.5 Convergence with Industry 6.0

The utmost demand arose from metal additive manufacturing technological progress to integrate IoT systems based on Industry 6.0 standards. Industry 6.0 goes further than Industry 4.0 by concentrating on automatic manufacturing processes which unite AI-based human collaboration and highly adaptive cyber-physical systems [30]. Better resilience and real-time adaptability for the IoT-enabled MAM system framework can be achieved by incorporating decentralized choices driven by the application of Edge AI. Digital twin interfaces, together with extended reality (XR) environments, allow human operators to work seamlessly in order to strengthen both efficiency and decrease cognitive load. The manufacturing industry expects this development to bring significant changes that will result in zero-defect production and improved sustainability aspects.

13.3 RECENT DEVELOPMENTS IN IOT-ENABLED MAM SYSTEMS

13.3.1 IoT-enhanced process control

Sophisticated IoT platforms that incorporate many sensor types (such as optical, acoustic, and thermal) for thorough process monitoring are becoming more and more common in modern MAM systems [31,32]. In SLM procedures, for example, real-time study of melt pool dynamics has been made possible by the combination of spectroscopic sensors and high-speed cameras [33]. These improvements enable operators to rapidly recognize process irregularities and alter parameters in real time, decreasing the occurrence of any defects during machining [34,35].

13.3.2 ML and predictive analytics

The potential of integrating IoT data streams with ML models for predictive analytics in MAM has been identified by recent studies [36]. These models are able to forecast when flaws will appear and accordingly suggest fixes before the part quality worsens [37]. In an investigation, convolutional neural networks (CNNs) were applied to in-situ sensor data to achieve high accuracy in defect prediction [38]. In another work, reinforcement learning algorithms were integrated with IoT frameworks to autonomously optimize process parameters, leading to a major advancement in the direction of completely autonomous MAM systems [39].

Figure 13.2 conceptualizes the real-time monitoring of process parameters (e.g., temperature, laser power) during an SLM build cycle. The plot indicates deviations from setpoints (dashed lines) and the corresponding corrective actions initiated by the control system [28,33].

13.3.3 Cyber-physical systems and digital twins

The concept of a digital twin, which makes a virtual duplicate of a physical process, has found its use in IoT-enabled MAM systems [40,41]. Manufacturers not only can simulate, but can also forecast the characteristics

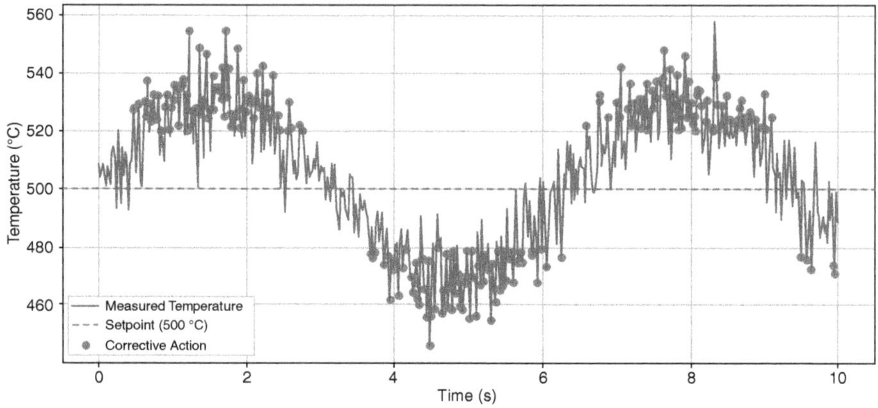

Figure 13.2 Conceptual plot of process parameter variations.

Table 13.1 Comparative analysis of IoT-enabled MAM processes

Process	*Key sensors/IoT components*	*Advantages*	*Limitations*	*References*
SLM	High-speed imaging, thermal cameras	High density, excellent surface finish	Sensitive to powder quality variations	[21,33,34,46–48]
LMD	Acoustic sensors, laser power sensors	Versatility in substrate repair and modification	High residual stresses	[25,49–51]
CSAM	Flow monitors, particle velocity sensors	Minimal thermal distortion	Limited to specific metal powders	[27,52–55]
Digital twin-enabled MAM	ML modules, IoT gateways, multi-sensor arrays	Real-time simulation and process optimization	High initial integration cost	[40,43,56–58]

of any MAM process under practical situations by constantly updating the digital twin with real-time sensor data [42]. Researchers have also demonstrated that digital twins are capable of drastically decreasing the time necessary for process optimization and troubleshooting [43,44]. Furthermore, cloud-based AM requires both secure and real-time standard communication protocols within the cloud platform for effective implementation [45].

A comparative overview of the primary IoT-enabled MAM processes is carried out in Table 13.1 based on the key sensors employed, their relative advantages, and limitations. Figure 13.3 can be referred to understand the core elements contributing to building an IoT-enabled MAM system.

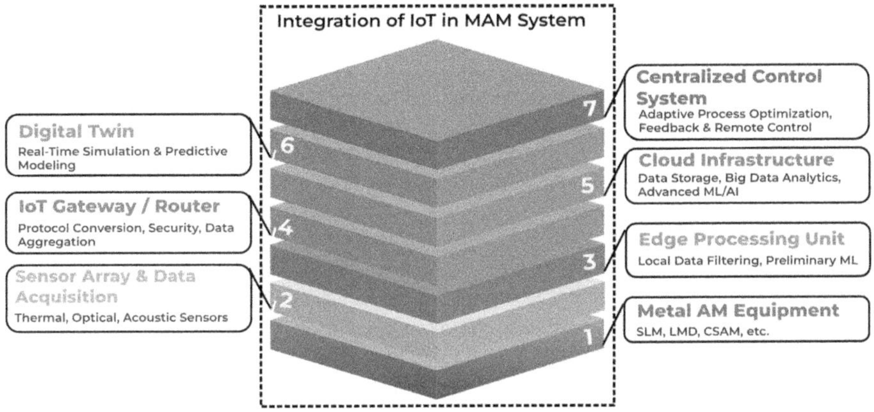

Figure 13.3 Representation of an IoT-integrated metal AM system [40,43,57].

13.4 IOT-METHODS, MATERIALS AND COMPARATIVE ANALYSIS

13.4.1 Essential IoT-methods and materials

Steady control of the process depends on multiple complex materials and tools used in IoT-enabled metal additive manufacturing systems. High-resolution sensors with thermal cameras and infrared sensors and high-speed imaging devices become vital in recording melt pool dynamics with temperature gradients [19,33]. The sensors connect with advanced data acquisition systems through wireless sensor networks supported by IoT gateways that provide real-time data transfer with secure communication channels [22,49]. Data analytics systems and machine learning algorithms provide predictive maintenance and quality monitoring using dependable technologies [36,39]. On being able to represent real-time simulations of physical processes, digital twin systems aid in process optimization and troubleshooting [40,43]. System protection and data confidentiality are achieved through the implementation of cybersecurity modules in the final stage [15,59].

Current research documents the main methods and sensors together with IoT tools employed in Table 13.2.

13.4.2 Comparative analysis

Research findings show that IoT enhancements deliver diverse outcomes for MAM processes regarding sensor connectivity, operational management, and system expansibility capabilities. Table 13.3 shows a summary of the advantages, together with limitations and potential growth areas for contemporary state-of-the-art systems.

Table 13.2 Methods and materials in IoT-enabled MAM

Category	*Tools/components*	*Functionality*	*References*
Sensors	Thermal cameras, infrared sensors, high-speed cameras	Monitor melt pool, temperature gradients, and process dynamics	[19,33,46]
Data acquisition systems	Wireless sensor networks, IoT gateways	Real-time data transmission from sensor nodes	[22,49]
Data analytics	ML algorithms (CNN, reinforcement learning)	Process optimization and defect prediction	[36,39]
Digital twin platforms	Simulation software, cloud computing	Virtual modeling and real-time simulation	[40,43]
Cybersecurity modules	Encryption protocols, secure IoT frameworks	Protect data integrity and ensure system security	[15,59]

Table 13.3 Analysis of IoT-enabled MAM systems

Aspect	*Strengths*	*Weaknesses*	*Opportunities for improvement*	*References*
Real-time monitoring	Immediate detection of anomalies	High data volume requires robust analytics	Advanced ML for rapid anomaly detection	[33,38,39,46]
Process control	Closed-loop feedback enables adaptive control	Dependence on sensor accuracy	Integration with digital twins	[34,40,43,47,48]
Data analytics	Predictive analytics enhances quality control	Complex data pre-processing	Cloud-based analytics for scalability	[36,39,40,56]
Cybersecurity	Emerging encryption techniques	Vulnerability to cyber-attacks	Standardized IoT security protocols	[15,59]
System scalability	Modular architectures allow integration	High initial setup cost	Cost-effective IoT solutions	[28,29,58,60]

13.5 DISCUSSION AND FUTURE RESEARCH

13.5.1 Discussion

There is still a significant research gap in terms of IoT protocol standardization and MAM cybersecurity measures. The existing systems use proprietary communication protocols that preclude different manufacturers and system entities from collaborating [15,61]. The fragmented industry structure impedes IoT solution scalability and introduces additional security risks that must be addressed [15,62]. The use of ML defect identification techniques in conjunction with process improvement approaches has potential [36,37], but their incorporation into standardized security frameworks requires further research [63,64]. Research into this gap can result in fully interoperable IoT-enabled MAM systems that are safe and scalable, hence advancing Industry 4.0 technology [65,66].

13.5.2 Future research directions

Future studies should focus on the following aspects:

- **Standardization of communication protocols:** The development of open standardized protocols for sensor data exchange helps multiple platforms and manufacturers to operate seamlessly across different systems [15,61,67].
- **Enhanced cybersecurity measures:** Implementation of robust encryption and real-time threat detection systems within IoT frameworks to protect sensitive process data [15,59,68].
- **Advanced AI integration:** In order to improve adaptive control and predictive maintenance, deep learning using reinforcement learning models is being investigated; nevertheless, secure integration techniques are needed [36,39,69].
- **Digital twin optimization:** Research on more complex digital twin systems that can accurately model complex process dynamics and instantly handle enormous amounts of data [40,43,70].
- **Cost-effective scalability:** Building affordable, flexible IoT solutions to guarantee that even small and medium-sized businesses can use these cutting-edge technologies [58,60,71].

The MAM community can overcome present barriers by using their expertise to establish an intelligent networked manufacturing system era [72–75].

13.6 DISCUSSION

Product quality and operational efficiency have increased as a result of the substantial advancements made in process monitoring, control, and predictive maintenance through the integration of IoT in MAM. With the help of a thorough examination of sensor technologies, digital twin integration, and data analytics approaches, this article has provided a comparative overview of the historical development and recent advancements in IoT-enabled MAM. Notwithstanding these developments, a major obstacle to creating completely interoperable and scalable systems is still the absence of defined IoT protocols and strong cybersecurity safeguards. For MAM to continue evolving under Industry 4.0, these issues must be addressed in future studies.

REFERENCES

1 Khan, Shahroz Akhtar, Pawan Kumar Arora, and Harish Kumar. "Retrospective Investigation on Emergence and Development of Additive Manufacturing." *Indian Journal of Engineering and Materials Sciences* 28, no. 02 (2021): 115–124. https://doi.org/10.56042/ijems.v28i2.36986

2 Kumar, Harish, Anuj Sharma, Yogesh Shrivastava, Shahroz Akhtar Khan, and Pawan Kumar Arora. "Parametric Influence of Process Parameters on the Wear Rate of 3D Printed Polylactic Acid Specimens." *Indian Journal of Pure & Applied Physics* 59, no. 3 (2021): 244–251. https://doi.org/10.56042/ijpap.v59i3.67760. Available at: https://nopr.niscpr.res.in/handle/123456789/56505.

3 Kumar, Aditya, Abhishek Choudhary, Abhishek Tiwari, Christopher James, Harish Kumar, Pawan Kumar Arora, and Shahroz Akhtar Khan. "An Investigation on Wear Characteristics of Additive Manufacturing Materials." *Materials Today: Proceedings* 47 (2021): 3654–60. doi:10.1016/j.matpr.2021.01.263.

4 Yadav, Abhishek, Adarsh Srivastav, Ankit Singh, M.D. Mushtaque, S.A. Khan, Harish Kumar, and P.K. Arora. "Investigation on the Materials Used in Additive Manufacturing: A Study." *Materials Today: Proceedings* 43 (2021): 154–57. doi:10.1016/j.matpr.2020.10.975.

5 Frazier, William E. "Metal Additive Manufacturing: A Review." *Journal of Materials Engineering and Performance* 23, no. 6 (April 8, 2014): 1917–28. doi:10.1007/s11665-014-0958-z.

6 Herzog, Dirk, Vanessa Seyda, Eric Wycisk, and Claus Emmelmann. "Additive Manufacturing of Metals." *Acta Materialia* 117 (September 2016): 371–92. doi:10.1016/j.actamat.2016.07.019.

7 Bikas, H., P. Stavropoulos, and G. Chryssolouris. "Additive Manufacturing Methods and Modelling Approaches: A Critical Review." *The International Journal of Advanced Manufacturing Technology* 83, nos. 1–4 (July 24, 2015): 389–405. doi:10.1007/s00170-015-7576-2.

8 Atzori, Luigi, Antonio Iera, and Giacomo Morabito. "The Internet of Things: A Survey." *Computer Networks* 54, no. 15 (October 2010): 2787–805. doi:10.1016/j.comnet.2010.05.010.

9 Borgia, Eleonora. "The Internet of Things Vision: Key Features, Applications and Open Issues." *Computer Communications* 54 (December 2014): 1–31. doi:10.1016/j.comcom.2014.09.008.

10 Zhang, Yikai, Shengnan Shen, Hui Li, and Yaowu Hu. "Review of in Situ and Real-Time Monitoring of Metal Additive Manufacturing Based on Image Processing." *The International Journal of Advanced Manufacturing Technology* 123, nos. 1–2 (September 29, 2022): 1–20. doi:10.1007/s00170-022-10178-3.

11 Grasso, Marco, and Bianca Maria Colosimo. "Process Defects and *in Situ* Monitoring Methods in Metal Powder Bed Fusion: A Review." *Measurement Science and Technology* 28, no. 4 (February 15, 2017): 044005. doi:10.1088/1361-6501/aa5c4f.

12 Inayathullah, Sirajudeen, and Raviteja Buddala. "Review of Machine Learning Applications in Additive Manufacturing." *Results in Engineering* 25 (March 2025): 103676. doi:10.1016/j.rineng.2024.103676.

13 Zubayer, Md Hasib, Chaoqun Zhang, and Yafei Wang. "Deep Learning-Based Automatic Defect Detection of Additive Manufactured Stainless Steel." *Metals* 13, no. 12 (December 7, 2023): 1987. doi:10.3390/met13121987.
14 Gupta, Shivam, Sonali Gupta, Kundan Kumar Pandey, Kritik Subodh Dwivedi, Shahroz Akhtar Khan, and P.K. Arora. "Influential Effects of Process Parameters of Fused Deposition Modelling on Wear of a PLA Specimen: A Comprehensive Review." *Smart Innovation, Systems and Technologies* (June 27, 2022): 579–90. doi:10.1007/978-981-19-0108-9_59.
15 Roman, Rodrigo, Jianying Zhou, and Javier Lopez. "On the Features and Challenges of Security and Privacy in Distributed Internet of Things." *Computer Networks* 57, no. 10 (July 2013): 2266–79. doi:10.1016/j.comnet.2012.12.018.
16 Gubbi, Jayavardhana, Rajkumar Buyya, Slaven Marusic, and Marimuthu Palaniswami. "Internet of Things (IOT): A Vision, Architectural Elements, and Future Directions." *Future Generation Computer Systems* 29, no. 7 (September 2013): 1645–60. doi:10.1016/j.future.2013.01.010.
17 Sharma, Ankita, Vibha Burman, and Shipra Aggarwal. "Role of IOT in Industry 4.0." *Lecture Notes in Electrical Engineering*, July 28, 2021, 517–28. doi:10.1007/978-981-16-1476-7_47.
18 Ghodsian, Nooshin, Khaled Benfriha, Adel Olabi, Varun Gopinath, and Aurélien Arnou. "Mobile Manipulators in Industry 4.0: A Review of Developments for Industrial Applications." *Sensors* 23, no. 19 (September 22, 2023): 8026. doi:10.3390/s23198026.
19 King, Wayne E., Holly D. Barth, Victor M. Castillo, Gilbert F. Gallegos, John W. Gibbs, Douglas E. Hahn, Chandrika Kamath, and Alexander M. Rubenchik. "Observation of Keyhole-Mode Laser Melting in Laser Powder-Bed Fusion Additive Manufacturing." *Journal of Materials Processing Technology* 214, no. 12 (December 2014): 2915–25. doi:10.1016/j.jmatprotec.2014.06.005.
20 Paul, Moses J., Qian Liu, James P. Best, Xiaopeng Li, Jamie J. Kruzic, Upadrasta Ramamurty, and Bernd Gludovatz. "Fracture Resistance of Alsi10mg Fabricated by Laser Powder Bed Fusion." *Acta Materialia* 211 (June 2021): 116869. doi:10.1016/j.actamat.2021.116869.
21 Altenburg, Simon J., Anne Straße, Andrey Gumenyuk, and Christiane Maierhofer. "In-Situ Monitoring of a Laser Metal Deposition (LMD) Process: Comparison of MWIR, Swir and High-Speed NIR Thermography." *Quantitative InfraRed Thermography Journal* 19, no. 2 (November 4, 2020): 97–114. doi:10.1080/17686733.2020.1829889.
22 Medina-García, Jonathan, Trinidad Sánchez-Rodríguez, Juan Galán, Aránzazu Delgado, Fernando Gómez-Bravo, and Raúl Jiménez. "A Wireless Sensor System for Real-Time Monitoring and Fault Detection of Motor Arrays." *Sensors* 17, no. 3 (February 25, 2017): 469. doi:10.3390/s17030469.
23 Fang, Qihang, Gang Xiong, MengChu Zhou, Tariku Sinshaw Tamir, Chao-Bo Yan, Huaiyu Wu, Zhen Shen, and Fei-Yue Wang. "Process Monitoring, Diagnosis and Control of Additive Manufacturing." *IEEE Transactions on Automation Science and Engineering* 21, no. 1 (January 2024): 1041–67. doi:10.1109/tase.2022.3215258.
24 Tartici, Idil, and Paulo Bartolo. "Exploring Materials Extrusion in 3D Printing: Real-Time Monitoring, and Machine Learning for Predicting Mechanical Properties." *Procedia CIRP* 130 (2024): 86–92. doi:10.1016/j.procir.2024.10.060.

25 Miller, David, Boyang Song, Michael Farnsworth, Divya Tiwari, Felicity Freeman, Iain Todd, and Ashutosh Tiwari. "IOT and Machine Learning for In-Situ Process Control Using Laser Based Additive Manufacturing (LBAM) Case Study." *Procedia CIRP* 104 (2021): 1813–18. doi:10.1016/j.procir.2021.11.306.
26 Wang, Yuanbin, Yuan Lin, Ray Y. Zhong, and Xun Xu. "IOT-Enabled Cloud-Based Additive Manufacturing Platform to Support Rapid Product Development." *International Journal of Production Research* 57, no. 12 (September 8, 2018): 3975–91. doi:10.1080/00207543.2018.1516905.
27 Lasi, Hermann, Peter Fettke, Hans-Georg Kemper, Thomas Feld, and Michael Hoffmann. "Industry 4.0." *Business & Information Systems Engineering* 6, no. 4 (2014): 239–42. https://doi.org/10.1007/s12599-014-0334-4
28 Monostori, László. "Cyber-Physical Production Systems: Roots, Expectations and R&D Challenges." *Procedia CIRP* 17 (2014): 9–13. doi:10.1016/j.procir.2014.03.115.
29 Lee, Edward A. "Cyber Physical Systems: Design Challenges." *2008 11th IEEE International Symposium on Object and Component-Oriented Real-Time Distributed Computing (ISORC)*, May 2008, 363–69. doi:10.1109/isorc.2008.25.
30 Lykov, Artem, Miguel Altamirano, Mikhail Konenkov, Valerii Serpiva, Koffivi Gbagbe, Ali Alabbas, Aleksey Fedoseev, Luis Moreno, Muhammad Khan, Ziang Guo, and Dzmitry Tsetserukou. "Industry 6.0: New Generation of Industry driven by Generative AI and Swarm of Heterogeneous Robots." (2024). doi:10.48550/arXiv.2409.10106.
31 Rana, Md Masud, and Narendra Dahotre. "IOT-Based Cyber-Physical Additive Manufacturing Systems: A Secure Communication Architecture, Research Challenges and Directions." *2021 6th International Conference on Inventive Computation Technologies (ICICT)*, January 20, 2021, 216–19. doi:10.1109/icict50816.2021.9358643.
32 Ehsan Toyserkani, Sarker Dyuti, Osezua Ibhadode, Farzad Liravi, Paola Russo, and Katayoon Taherkhani. "Monitoring and Quality Assurance for Metal Additive Manufacturing." *Metal Additive Manufacturing*, October 22, 2021, 507–75. doi:10.1002/9781119210801.ch11.
33 Wang, Rongxuan, David Garcia, Rakesh R. Kamath, Chaoran Dou, Xiaohan Ma, Bo Shen, Hahn Choo, Kamel Fezzaa, Hang Z. Yu, and Zhenyu Kong. "In Situ Melt Pool Measurements for Laser Powder Bed Fusion Using Multi Sensing and Correlation Analysis." *Scientific Reports* 12, no. 1 (August 12, 2022). doi:10.1038/s41598-022-18096-w.
34 Wang, Rongxuan, Benjamin Standfield, Chaoran Dou, Andrew Chung Law, and Zhenyu Kong. "Real-Time Process Monitoring and Closed-Loop Control on Laser Power via a Customized Laser Powder Bed Fusion Platform." *SSRN Electronic Journal* (2022). doi:10.2139/ssrn.4187431.
35 Chen, Lequn, Guijun Bi, Xiling Yao, Jinlong Su, Chaolin Tan, Wenhe Feng, Michalis Benakis, Youxiang Chew, and Seung Ki Moon. "In-Situ Process Monitoring and Adaptive Quality Enhancement in Laser Additive Manufacturing: A Critical Review." *Journal of Manufacturing Systems* 74 (June 2024): 527–74. doi:10.1016/j.jmsy.2024.04.013.

36 Zhang, Xijun, Dianming Chu, Xinyue Zhao, Chenyu Gao, Lingxiao Lu, Yan He, and Wenjuan Bai. "Machine Learning-Driven 3D Printing: A Review." *Applied Materials Today* 39 (August 2024): 102306. doi:10.1016/j.apmt.2024.102306.

37 Cheepu, Muralimohan. "Machine Learning Approach for the Prediction of Defect Characteristics in Wire Arc Additive Manufacturing." *Transactions of the Indian Institute of Metals* 76, no. 2 (August 13, 2022): 447–55. doi:10.1007/s12666-022-02715-1.

38 Kozhay, Kaisar, Shynggys Turarbek, Tolkyn Asselbekova, Md. Hazrat Ali, and Essam Shehab. "Convolutional Neural Network-Based Defect Detection Technique in FDM Technology." *Procedia Computer Science* 231 (2024): 119–28. doi:10.1016/j.procs.2023.12.183.

39 Dharmadhikari, Susheel, Nandana Menon, and Amrita Basak. "A Reinforcement Learning Approach for Process Parameter Optimization in Additive Manufacturing." *Additive Manufacturing* 71 (June 2023): 103556. doi:10.1016/j.addma.2023.103556.

40 Tao, Fei, He Zhang, Ang Liu, and A. Y. Nee. "Digital Twin in Industry: State-of-the-Art." *IEEE Transactions on Industrial Informatics* 15, no. 4 (April 2019): 2405–15. doi:10.1109/tii.2018.2873186.

41 Kritzinger, Werner, Matthias Karner, Georg Traar, Jan Henjes, and Wilfried Sihn. "Digital Twin In Manufacturing: A Categorical Literature Review and Classification." *IFAC-PapersOnLine* 51, no. 11 (2018): 1016–22. doi:10.1016/j.ifacol.2018.08.474.

42 Li, Jingdong, Xiaochen Wang, Quan Yang, Youzhao Sun, Jianwei Zhao, Xing Mao, and Haotang Qie. "IOT-Based Framework for Digital Twins in Steel Production: A Case Study of Key Parameter Prediction and Optimization for CSR." *Expert Systems with Applications* 250 (September 2024): 123909. doi:10.1016/j.eswa.2024.123909.

43 Stavropoulos, Panagiotis, Alexios Papacharalampopoulos, and Christos K. Michail. "Digital Twin-Driven Multi-Variable Process Control of Thermal Manufacturing Processes." *Procedia CIRP* 107 (2022): 752–57. doi:10.1016/j.procir.2022.05.057.

44 Ben Amor, Sabrine, Nessrine Elloumi, Ameni Eltaief, Borhen Louhichi, Nashmi H. Alrasheedi, and Abdennour Seibi. "Digital Twin Implementation in Additive Manufacturing: A Comprehensive Review." *Processes* 12, no. 6 (May 22, 2024): 1062. doi:10.3390/pr12061062.

45 Shahroz Akhtar Khan, Ranjan Arora, Harish Kumar, and Pawan Kumar Arora. "A Perspective on Advances in Cloud-Based Additive Manufacturing." *Evergreen* 9, no. 3 (September 2022): 861–69. doi:10.5109/4843119.

46 Zhao, Cang, Kamel Fezzaa, Ross W. Cunningham, Haidan Wen, Francesco De Carlo, Lianyi Chen, Anthony D. Rollett, and Tao Sun. "Real-Time Monitoring of Laser Powder Bed Fusion Process Using High-Speed X-Ray Imaging and Diffraction." *Scientific Reports* 7, no. 1 (June 15, 2017). doi:10.1038/s41598-017-03761-2.

47 Eshawish, Naeem, Savko Malinov, Wei Sha, and Patrick Walls. "Microstructure and Mechanical Properties of TI-6AL-4V Manufactured by Selective Laser Melting after Stress Relieving, Hot Isostatic Pressing Treatment, and Post-Heat Treatment." *Journal of Materials Engineering and Performance* 30, no. 7 (April 15, 2021): 5290–96. doi:10.1007/s11665-021-05753-w.
48 Bonesso, Massimiliano, Pietro Rebesan, Claudio Gennari, Simone Mancin, Razvan Dima, Adriano Pepato, and Irene Calliari. "Effect of Particle Size Distribution on Laser Powder Bed Fusion Manufacturability of Copper." *BHM Berg- Und Hüttenmännische Monatshefte* 166, no. 5 (April 21, 2021): 256–62. doi:10.1007/s00501-021-01107-0.
49 Gaja, Haythem, and Frank Liou. "Defects Monitoring of Laser Metal Deposition Using Acoustic Emission Sensor." *The International Journal of Advanced Manufacturing Technology* 90, nos. 1–4 (September 3, 2016): 561–74. doi:10.1007/s00170-016-9366-x.
50 Kumar, L. Jyothish, and C.G. Krishnadas Nair. "Laser Metal Deposition Repair Applications for Inconel 718 Alloy." *Materials Today: Proceedings* 4, no. 10 (2017): 11068–77. doi:10.1016/j.matpr.2017.08.068.
51 Caiazzo, Fabrizia, Vittorio Alfieri, and Giovanni Bolelli. "Residual Stress in Laser-Based Directed Energy Deposition of Aluminum Alloy 2024: Simulation and Validation." *The International Journal of Advanced Manufacturing Technology* 118, nos. 3–4 (September 14, 2021): 1197–211. doi:10.1007/s00170-021-07988-2.
52 Zou, Yu. "Cold Spray Additive Manufacturing: Microstructure Evolution and Bonding Features." *Accounts of Materials Research* 2, no. 11 (October 12, 2021): 1071–81. doi:10.1021/accountsmr.1c00138.
53 Arkhipov, Ivan, Uğur Kokal, and Ozan Özdemir. "Aeroacoustic Process Monitoring and Anomaly Detection in Cold Spray Additive Manufacturing." *Journal of Thermal Spray Technology* 34, no. 1 (February 2025): 97–119. doi:10.1007/s11666-025-01934-4.
54 Prashar, Gaurav, and Hitesh Vasudev. "A Comprehensive Review on Sustainable Cold Spray Additive Manufacturing: State of the Art, Challenges and Future Challenges." *Journal of Cleaner Production* 310 (August 2021): 127606. doi:10.1016/j.jclepro.2021.127606.
55 Yin, Shuo, Ningsong Fan, Chunjie Huang, Yingchun Xie, Chao Zhang, Rocco Lupoi, and Wenya Li. "Towards High-Strength Cold Spray Additive Manufactured Metals: Methods, Mechanisms, and Properties." *Journal of Materials Science & Technology* 170 (January 2024): 47–64. doi:10.1016/j.jmst.2023.05.047.
56 Lu, Yuqian, Chao Liu, Kevin I-Kai Wang, Huiyue Huang, and Xun Xu. "Digital Twin-Driven Smart Manufacturing: Connotation, Reference Model, Applications and Research Issues." *Robotics and Computer-Integrated Manufacturing* 61 (February 2020): 101837. doi:10.1016/j.rcim.2019.101837.
57 Negri, Elisa, Luca Fumagalli, and Marco Macchi. "A Review of the Roles of Digital Twin in CPS-Based Production Systems." *Procedia Manufacturing* 11 (2017): 939–48. doi:10.1016/j.promfg.2017.07.198.
58 Malik, Ali Ahmad. "The Economic Impact of Digital Twin Technology on Manufacturing Systems." *2024 Winter Simulation Conference (WSC)*, December 15, 2024, 1623–33. doi:10.1109/wsc63780.2024.10838771.

59 Sicari, S., A. Rizzardi, L.A. Grieco, and A. Coen-Porisini. "Security, Privacy and Trust in Internet of Things: The Road Ahead." *Computer Networks* 76 (January 2015): 146–64. doi:10.1016/j.comnet.2014.11.008.

60 Roy, Marc-Antoine, and Georges Abdul-Nour. "Integrating Modular Design Concepts for Enhanced Efficiency in Digital and Sustainable Manufacturing: A Literature Review." *Applied Sciences* 14, no. 11 (May 25, 2024): 4539. doi:10.3390/app14114539.

61 Banafa, Ahmed. "IOT Standardization and Implementation Challenges." *Introduction to Internet of Things (IoT)*, May 22, 2023, 45–50. doi:10.1201/9781003426240-8.

62 Wang, Maoli, Yu Sun, Hongtao Sun, and Bowen Zhang. "Security Issues on Industrial Internet of Things: Overview and Challenges." *Computers* 12, no. 12 (December 8, 2023): 256. doi:10.3390/computers12120256.

63 Rehman, Zubaida, Noshina Tariq, Syed Atif Moqurrab, Joon Yoo, and Gautam Srivastava. "Machine Learning and Internet of Things Applications in Enterprise Architectures: Solutions, Challenges, and Open Issues." *Expert Systems* 41, no. 1 (October 18, 2023). doi:10.1111/exsy.13467.

64 Dmitrieva, Ekaterina, Gopal Krishna, Soosan Chhabra, Avvari Pavithra, and Kshama Sharma. "IOT and AI Integration: An Experiment on Smart Manufacturing Efficiency in Industry 5.0." *BIO Web of Conferences* 86 (2024): 01062. doi:10.1051/bioconf/20248601062.

65 Oks, Sascha Julian, Max Jalowski, Michael Lechner, Stefan Mirschberger, Marion Merklein, Birgit Vogel-Heuser, and Kathrin M. Möslein. "Cyber-Physical Systems in the Context of Industry 4.0: A Review, Categorization and Outlook." *Information Systems Frontiers* 26, no. 5 (April 7, 2022): 1731–72. doi:10.1007/s10796-022-10252-x.

66 Zeid, Abe, Sarvesh Sundaram, Mohsen Moghaddam, Sagar Kamarthi, and Tucker Marion. "Interoperability in Smart Manufacturing: Research Challenges." *Machines* 7, no. 2 (April 2, 2019): 21. doi:10.3390/machines7020021.

67 OPC Foundation. (2018). *OPC Unified Architecture Specification*. Phoenix, AZ: OPC Foundation.

68 Miorandi, Daniele, Sabrina Sicari, Francesco De Pellegrini, and Imrich Chlamtac. "Internet of Things: Vision, Applications and Research Challenges." *Ad Hoc Networks* 10, no. 7 (September 2012): 1497–516. doi:10.1016/j.adhoc.2012.02.016.

69 Khdoudi, Abdelmoula, Tawfik Masrour, Ibtissam El Hassani, and Choumicha El Mazgualdi. "A Deep-Reinforcement-Learning-Based Digital Twin for Manufacturing Process Optimization." *Systems* 12, no. 2 (January 24, 2024): 38. doi:10.3390/systems12020038.

70 Boschert, Stefan, and Roland Rosen. "Digital Twin—the Simulation Aspect." *Mechatronic Futures* (2016): 59–74. doi:10.1007/978-3-319-32156-1_5.

71 Porter, M. E., and J. E. Heppelmann. "How Smart, Connected Products Are Transforming Competition." *Harvard Business Review* 92, no. 11 (2014): 64–88.

72 Ashima, Reem, Abid Haleem, Mohd Javaid, and Shanay Rab. "Understanding the Role and Capabilities of Internet of Things-Enabled Additive Manufacturing through its Application Areas." *Advanced Industrial and Engineering Polymer Research* 5, no. 3 (July 2022): 137–42. doi:10.1016/j.aiepr.2021.12.001.
73 O'Connell, Eoin, William O'Brien, Mangolika Bhattacharya, Denis Moore, and Mihai Penica. "Digital Twins: Enabling Interoperability in Smart Manufacturing Networks." *Telecom* 4, no. 2 (May 11, 2023): 265–78. doi:10.3390/telecom4020016.
74 Soori, Mohsen, Behrooz Arezoo, and Roza Dastres. "Digital Twin for Smart Manufacturing, a Review." *Sustainable Manufacturing and Service Economics* 2 (April 2023): 100017. doi:10.1016/j.smse.2023.100017.
75 Phuyal, Sudip, Diwakar Bista, and Rabindra Bista. "Challenges, Opportunities and Future Directions of Smart Manufacturing: A State of Art Review." *Sustainable Futures* 2 (2020): 100023. doi:10.1016/j.sftr.2020.100023.

Index

www.ingramcontent.com/pod-product-compliance
Ingram Content Group UK Ltd.
Pitfield, Milton Keynes, MK11 3LW, UK
UKHW022224190926
475702UK00003B/489
* 9 7 8 1 0 3 2 9 9 9 8 8 3 *